Hodder Gibson
Scottish Examination Materials

INTERMEDIATE 2 PHYSICS

Revision Notes & Questions

Arthur Baillie *and* Andrew McCormick

Editor: Rothwell Glen

Hodder Gibson
A MEMBER OF THE HODDER HEADLINE GROUP

Orders: please contact Bookpoint Ltd, 130 Milton Park, Abingdon, Oxon OX14 4SB. Telephone: (44) 01235 827720. Fax: (44) 01235 400454. Lines are open 9.00–5.00, Monday to Saturday, with a 24-hour message answering service. Visit our website at www.hoddereducation.co.uk. Hodder Gibson can be contacted direct on: Tel: 0141 848 1609; Fax: 0141 889 6315; email: hoddergibson@hodder.co.uk

First published in 2007 by
Hodder Gibson, an imprint of Hodder Education
and a member of the Hodder Headline Group,
an Hachette Livre UK company,
2a Christie Street
Paisley PA1 1NB

Impression number 5 4 3 2 1
Year 2010 2009 2008 2007

Cover photo TEK Image/Science Photo Library
Illustrations by Fakenham Photosetting Limited
Typeset in 10.5pt Garamond by Fakenham Photosetting Limited
Printed in Great Britain by Martins The Printers, Berwick-upon-Tweed.

A catalogue record for this title is available from the British Library

ISBN-13: 978-0-340-94012-9

CONTENTS

INTRODUCTION

'But in physics I soon learned to scent out the paths that led to the depths, and to disregard everything else, all the many things that clutter up the mind, and divert it from the essential. The hitch in this was, of course, the fact that one had to cram all this stuff into one's mind for the examination, whether one liked it or not.'

Albert Einstein

'All of physics is either impossible or trivial. It is impossible until you understand it, and then it becomes trivial.'

Ernest Rutherford (1st Baron Rutherford of Nelson) (1871–1937) Nobel Prize winner

This book is written as a companion to our second edition of Intermediate 2 Physics. It can be used to improve the skills which are necessary in a numerical problem solving context and to explain and describe a variety of physical phenomena relevant to the course. We are indebted to Rothwell Glen for the meticulous care and attention which he has given to the various questions and demands asked of him. He has pointed out omissions and clarified many of our ideas. We are also grateful to our families for the time they have allowed us to compile this book. It is our hope that students will see beyond the calculations to the real depth and clarity which is the wonder of physics.

Andrew McCormick, Arthur Baillie and Rothwell Glen 2007

Chapter 1

ELECTRICITY AND ELECTRONICS

1.1 Circuits

- Materials which allow electrons to move through them easily are called conductors. Conductors are mainly metals.
- Materials which do not allow electrons to move through them easily are called insulators. Examples of insulators are glass, plastic, wood and air.
- Charge is measured in coulombs (C), current in amperes (A) and time in seconds (s).
- Charge transferred = current × time, i.e. $Q = It$.
- The voltage of a supply is a measure of the energy given to one coulomb of charge passing through the supply.
- An ammeter is connected in series with a component.
- A voltmeter is connected in parallel across a component.
- Voltage or potential difference (p.d.) is measured in volts (V), current in amperes (A) and resistance in ohms (Ω).
- Voltage (p.d.) across a resistor = current in resistor × resistance of resistor, i.e. $V = IR$. This is known as Ohm's Law.
- The resistance of a resistor remains constant for different currents provided the temperature of the resistor does not change.
- In a series circuit:

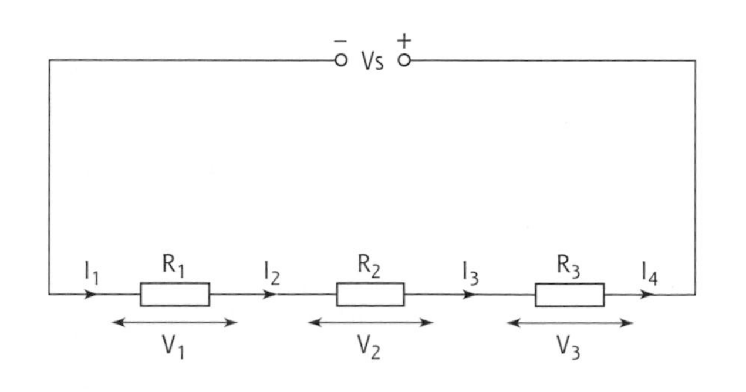

  - the current is the same at all points, i.e. $I_1 = I_2 = I_3 = I_4$
  - the supply voltage is equal to the sum of the voltages (p.d.s) across components, i.e. $V_S = V_1 + V_2 + V_3$

- the total (or combined) resistance is found using $R_T = R_1 + R_2 + R_3$
- the total resistance is greater than the value of the largest resistance connected in series.

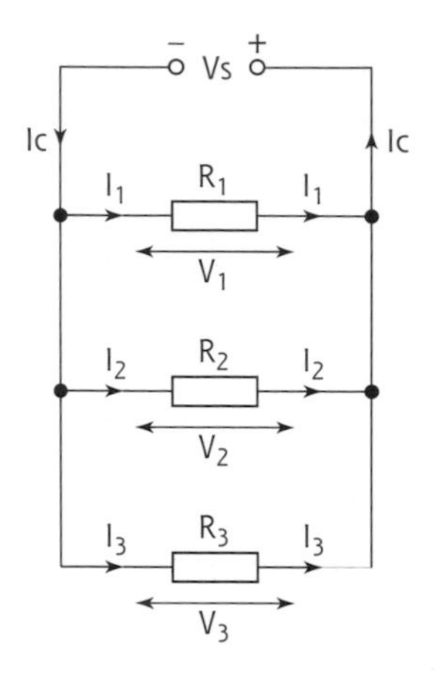

❍ In a parallel circuit:

- the circuit current is equal to the sum of the currents in the branches, i.e. $I_{circuit} = I_1 + I_2 + I_3$
- the voltage (p.d.) across components is the same, i.e. $V_S = V_1 = V_2 = V_3$
- the total (or combined) resistance is found by $$\frac{1}{R_T} = \frac{1}{R_1} + \frac{1}{R_2} + \frac{1}{R_3}$$
- the total resistance is less than the value of the smallest resistance connected in parallel.

❍ In a parallel circuit containing identical resistances

$$R_T = \frac{\text{resistance of one resistor}}{\text{number of resistors}}$$

QUESTIONS

1 Look at the diagram.

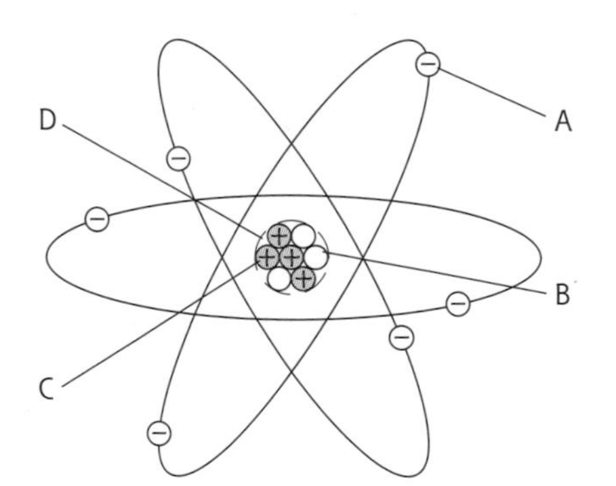

Match each of the letters A, B, C and D with the correct word from the box.

electron neutron nucleus proton

2 Draw up a table with the headings 'conductor' and 'insulator'. Complete the table by placing the words in the box into the correct column.

glass air aluminium foil 2p coin iron nail metal ruler paper plastic pen wooden pencil

> **Example question**
>
> The current in a wire is 0.30 A. Calculate the charge transferred through the wire in two minutes.
>
> *Solution*
>
> $Q = It = 0.30 \times (2 \times 60) = 36\,C$

3 In the table shown, calculate the value of each missing quantity.

	Charge (C)	Current (A)	Time (s)
a)		6.0	20
b)		0.40	5
c)	200		4
d)	36		180
e)	25	0.50	
f)	550	2.5	

4 Calculate the charge transferred in each case when there is:

a) a current of 4.0 A in a resistor for a time of 25 s

b) a current of 2.0 A in a motor for a time of three minutes

c) a current of 5.0 mA in a lamp for a time of two hours.

5 A hairdryer is switched on for two minutes. There is a current of 5.0 A in the element of the hairdryer. Calculate the charge transferred in this time.

6 Calculate the time taken for a current of 1.5 A to transfer 120 C of charge.

7 The charge transferred by a lamp in eight minutes is 1920 C. Calculate the current in the lamp.

8 In the following sentences the words represented by the letters A, B, C, D and E are missing.

A conductor is a material in which ___A___ are free to move. When electrons (negative charges) move in one direction in an electrical circuit a ___B___ is formed. Current is measured in ___C___.

The voltage of a supply is a measure of the ___D___ given to the charges (electrons) in an electrical circuit. Voltage or potential difference is measured in ___E___.

Match each letter with the correct word from the box.

amperes current electrons energy neutrons protons voltage volts

9 A games console works from a 5 V supply. What does a supply voltage of 5 V mean?

10 Draw the circuit symbol for:

a) a battery

b) a lamp

c) a switch

d) a resistor

e) a variable resistor

f) a fuse

g) an ammeter

h) a voltmeter.

11 A pupil draws the circuit shown.

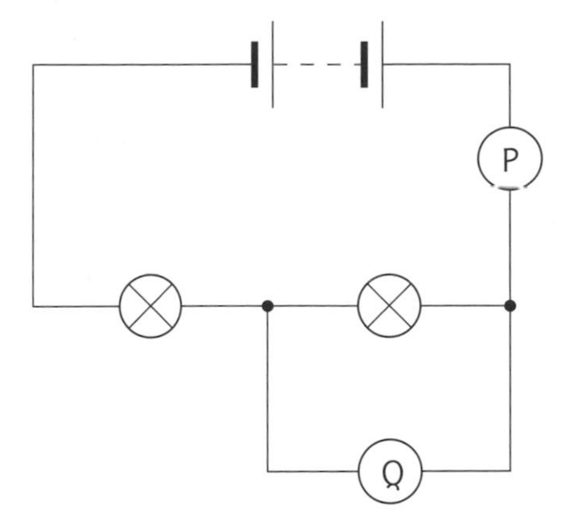

Name the type of meter labelled:

a) P

b) Q.

12 Redraw each of the diagrams shown.

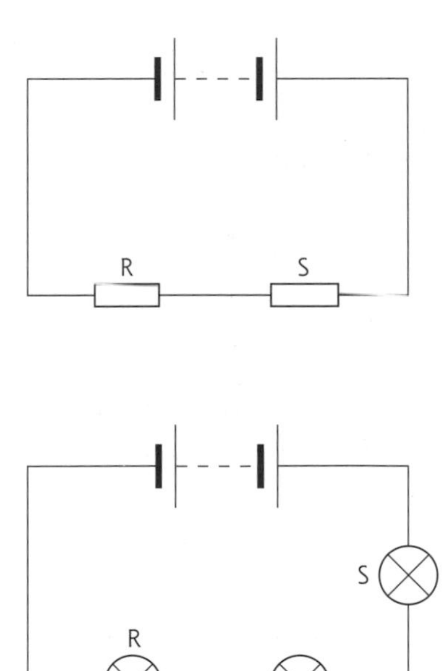

On each diagram show how:

a) a voltmeter is used to measure the voltage across component R

b) an ammeter is used to measure the current in component S.

13 In the following sentences the words represented by the letters A, B, C and D are missing.

The opposition to the flow of current is called ___A___. Resistance is measured in ___B___. Increasing the resistance of a circuit, ___C___ the current in the circuit.

The equation $V = IR$ is known as Ohm's Law. The ratio of $\frac{V}{I}$ for a resistor remains approximately ___D___ for different currents.

Match each letter with the correct word from the box.

amperes constant decreases increases joules ohms resistance volts

Example question

When switched on, the element of a kettle has a resistance of 23 Ω. The current in the element is 10 A. Calculate the potential difference across the ends of the element.

Solution

$V = IR = 10 \times 23 = 230\,\text{V}$

14 In the table shown, calculate the value of each missing quantity.

	Potential difference (V)	Current (A)	Resistance (Ω)
a)		1.0	10
b)		4.0	3
c)	5.0		100
d)	230		40
e)	0.02	0.01	
f)	1000	2.5	

15 Three circuits are set up as shown.

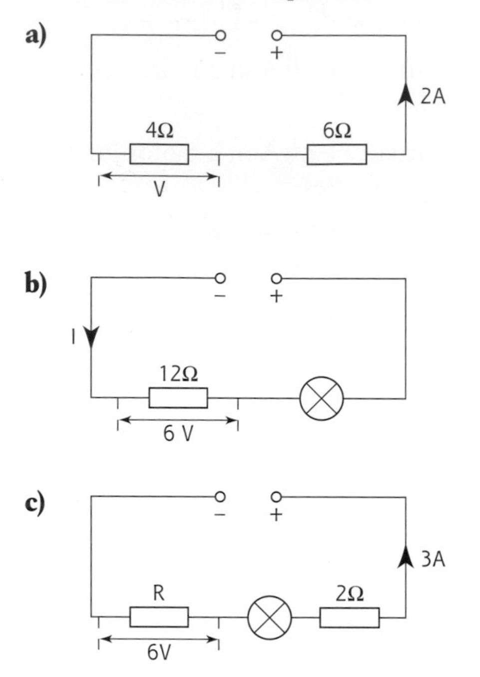

Find the values of V, I, or R in the circuits.

16 There is a current of 0.5 A in an 8 Ω resistor. Calculate the potential difference across the resistor.

17 A range of electrical appliances are to be connected to the 230 V mains using a flex. The maximum current in the flex is 13 A.

What is the smallest resistance of an electrical appliance which can be safely connected to the 230 V mains using this flex?

18 A lamp is rated at 48 W, 4 A. The lamp is connected to a 12 V supply. Calculate the resistance of the lamp when it is working at its correct rating.

19 Circuits are set up as shown.

a)

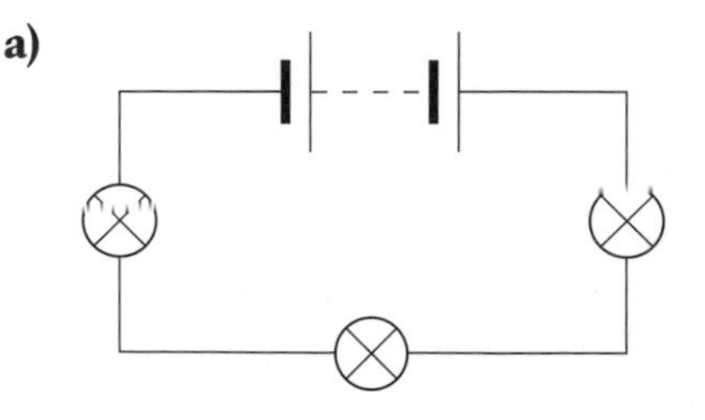

b)

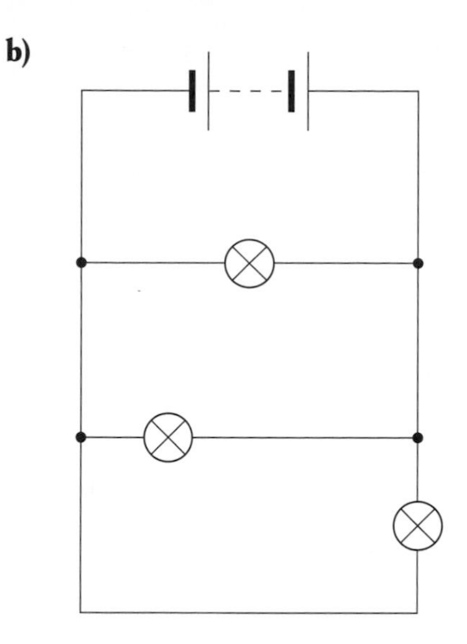

c)

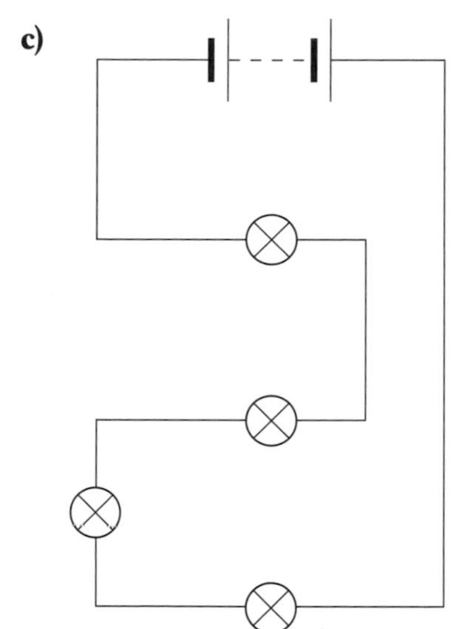

d)

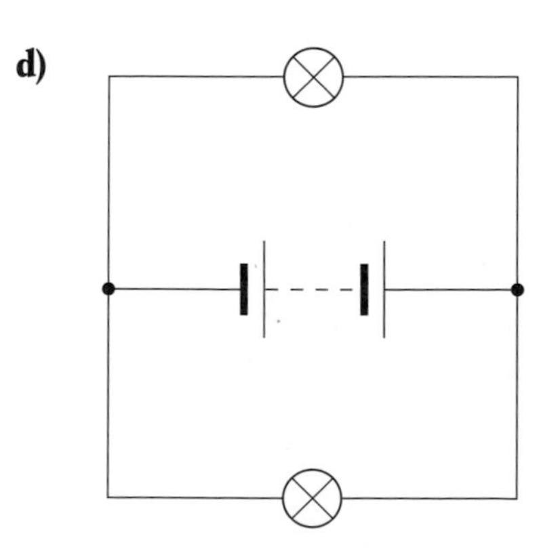

For each circuit state whether the lamps are connected in series or parallel.

20 You are given a circuit in which the various components are connected in series.

a) How does the current drawn from the supply compare with the current in each of the components?

b) How does the voltage of the supply compare with the voltage across each of the components?

21 You are given a circuit in which the various components are connected in parallel.

a) How does the current drawn from the supply compare with the current in each of the components?

b) How does the voltage across the supply compare with the voltage across each of the components?

22 Circuits are set up as shown.

a)

b)

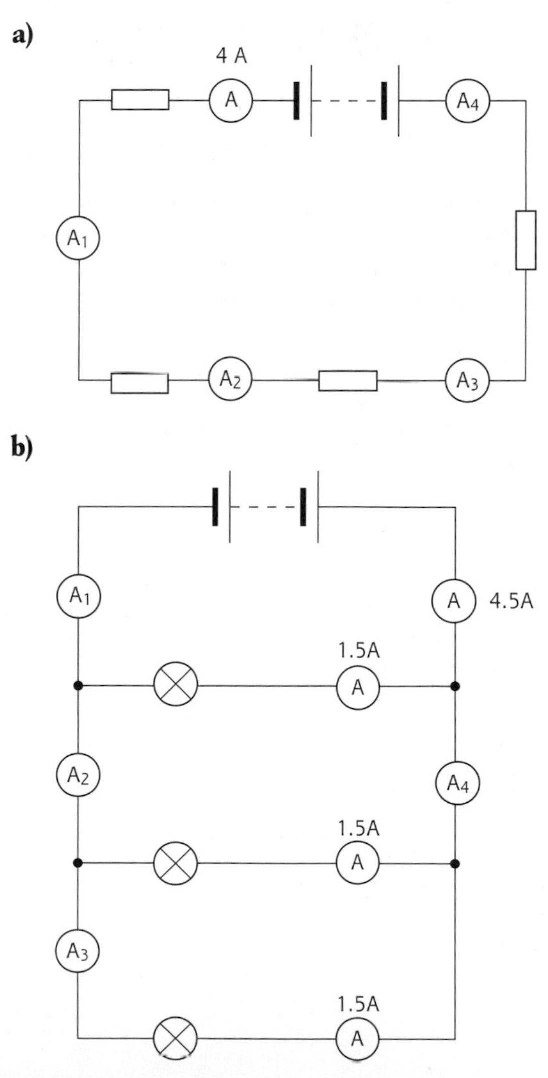

c)

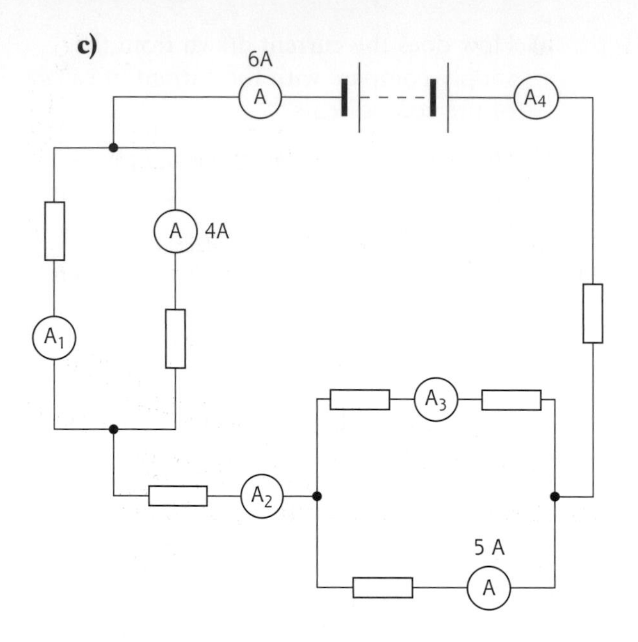

What are the readings on ammeters A_1, A_2, A_3 and A_4 in each circuit?

23 Circuits are set up as shown.

a)

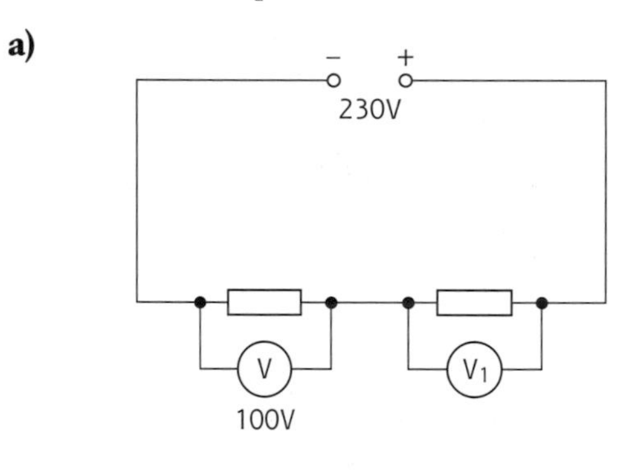

b)

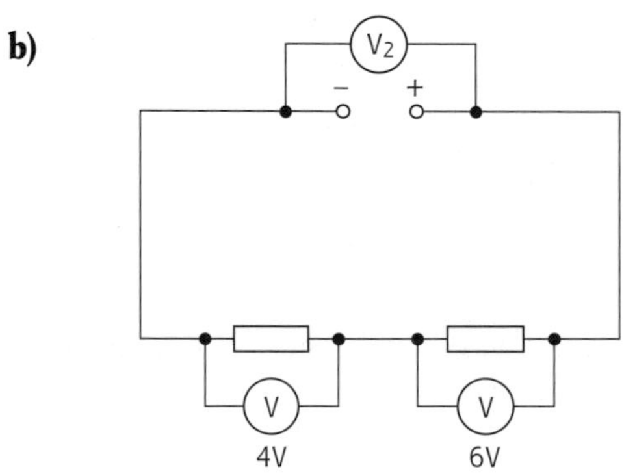

c)

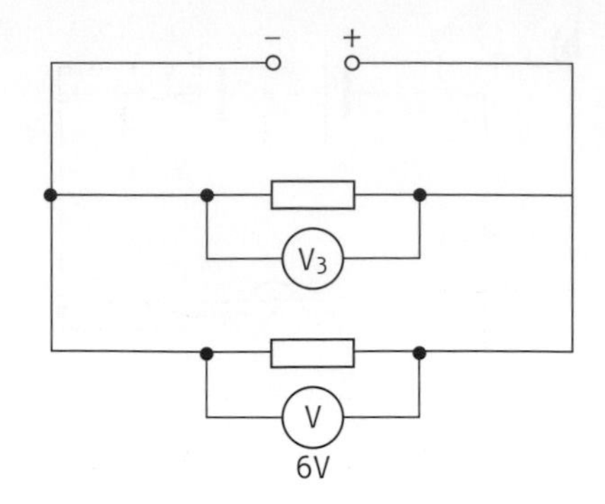

d)

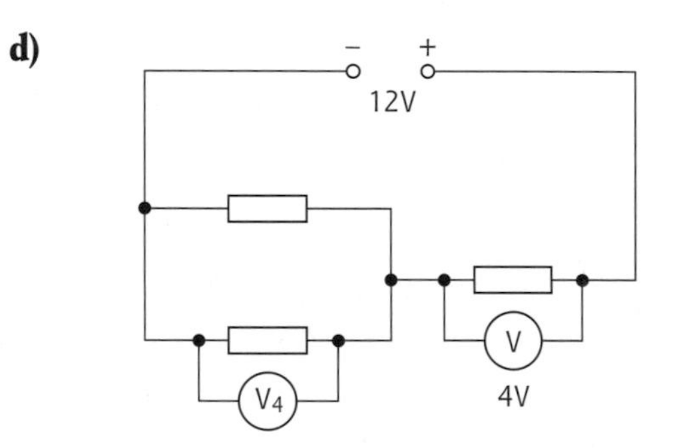

What are the readings on voltmeters V_1, V_2, V_3 and V_4?

24 Circuits are set up as shown.

a)

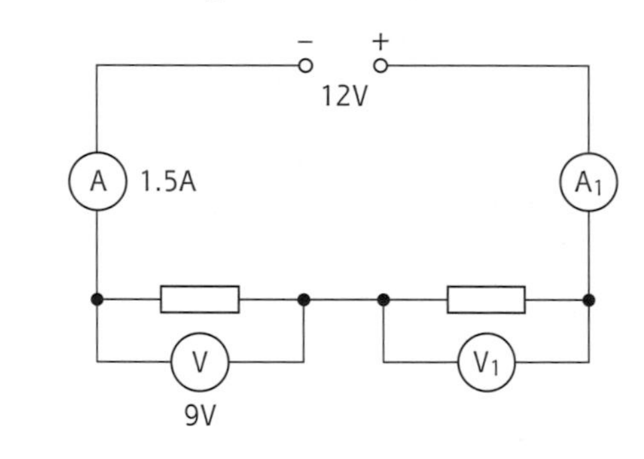

b)

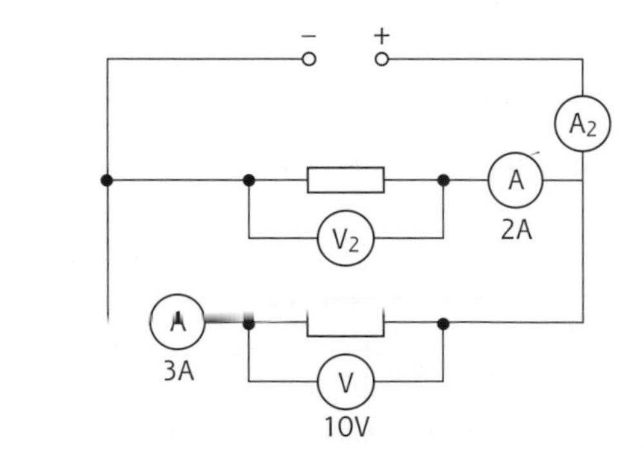

c)

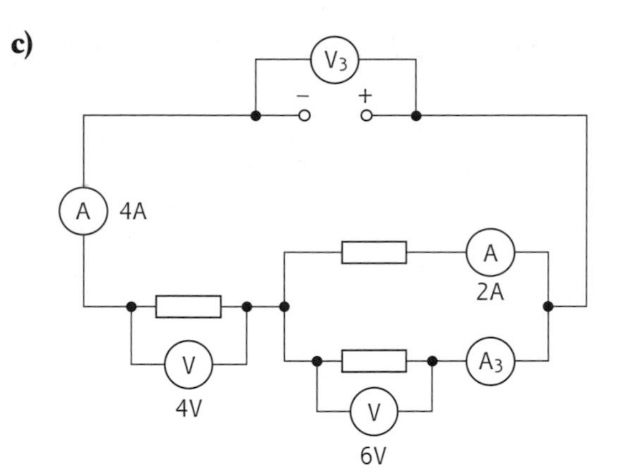

What are the readings on meters A_1, A_2 and A_3, and V_1, V_2, and V_3?

Example question

Three resistors of value 10 Ω, 12 Ω and 33 Ω are connected in series. Calculate the total resistance of the resistors.

Solution

$R_T = R_1 + R_2 + R_3 = 10 + 12 + 33 = 55\,\Omega$

Example question

Two resistors of value 30 Ω and 60 Ω are connected in parallel. Calculate the total resistance of the resistors.

Solution

$$\frac{1}{R_T} = \frac{1}{R_1} + \frac{1}{R_2} = \frac{1}{30} + \frac{1}{60}$$

$$= 0.033 + 0.017 = 0.05$$

$$\frac{1}{R_T} = 0.05$$

$$R_T = \frac{1}{0.05} = 20\,\Omega$$

25 Resistor networks are set up as shown.

a)

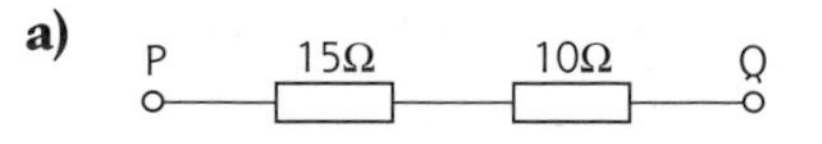

b)

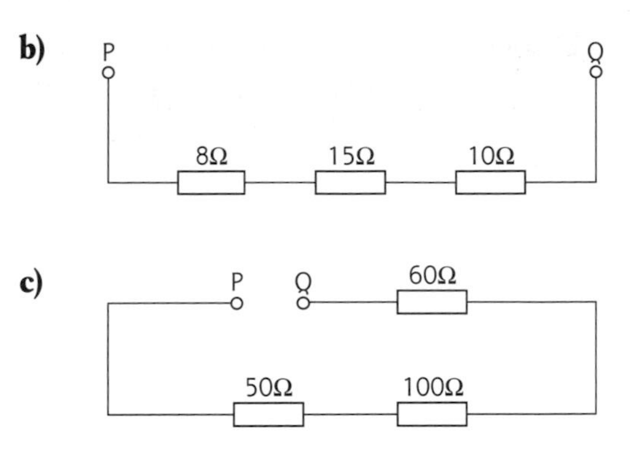

c)

Calculate the resistance between P and Q in each network.

26 Resistor networks are set up as shown.

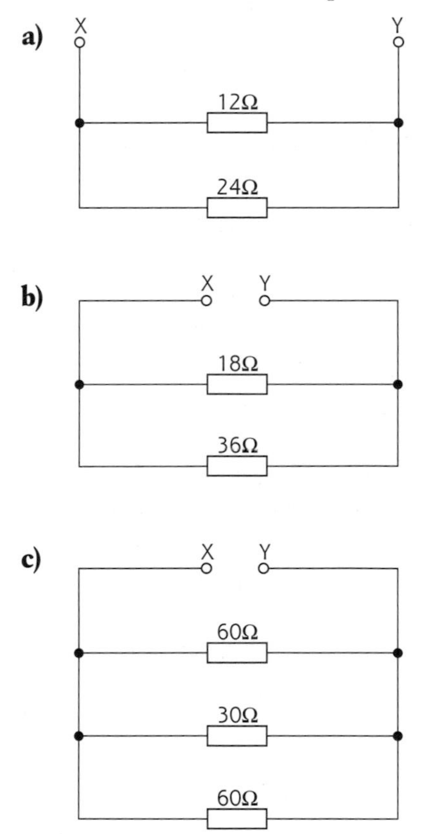

Calculate the resistance between X and Y in each network.

27 Resistor networks are set up as shown.

a)

b)

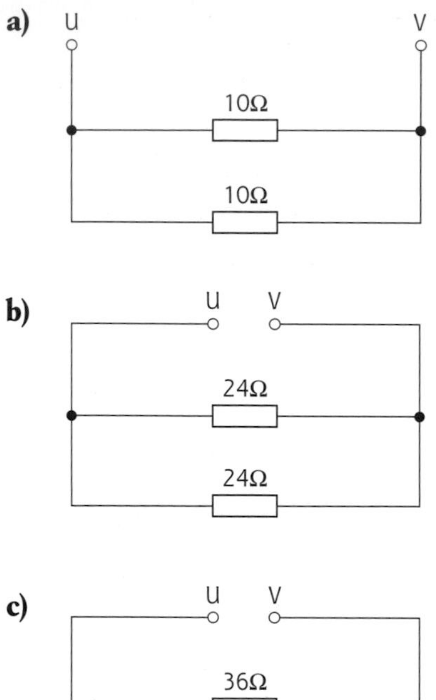

c)

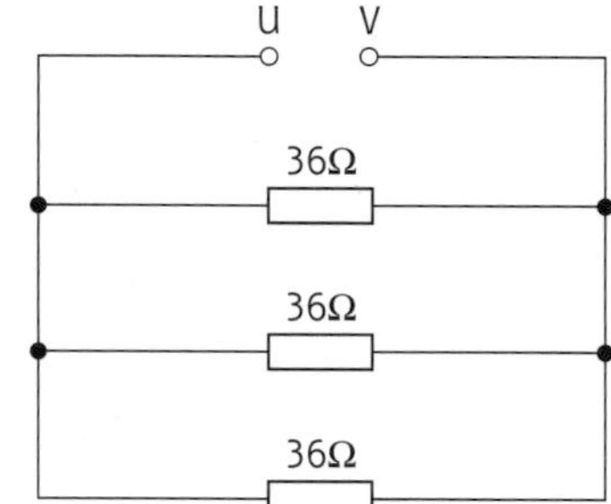

Calculate the resistance between U and V in each network.

28 Resistor networks are set up as shown.

a)

b)

c)

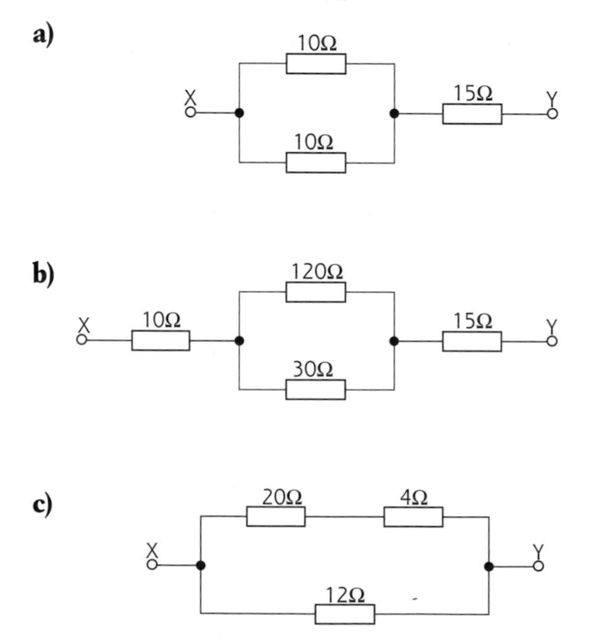

Calculate the resistance between X and Y in each network.

29 In the following sentence the word represented by the letter A is missing.

A potential divider either consists of two or more resistors connected in ___A___, or a variable resistor, across a supply.

Match letter A with the correct word from the box.

parallel series

Example question

A circuit is set up as shown

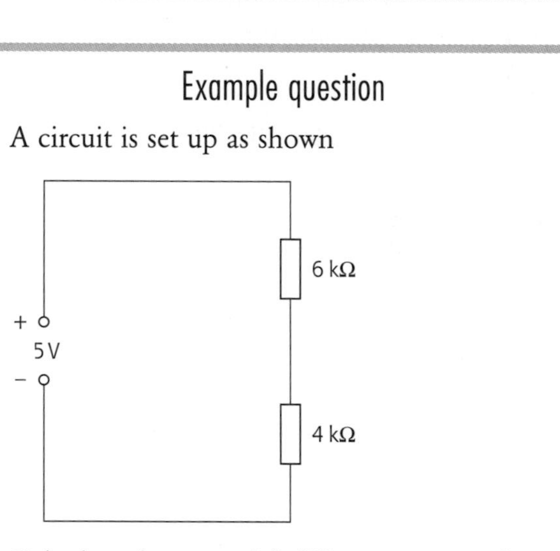

Calculate the potential difference across the 4 kΩ resistor.

Solution

$R_T = R_1 + R_2 = 4000 + 6000 = 10\,000\,\Omega$

$V_S = I_{circuit} \times R_T$

$5 = I_{circuit} \times 10\,000$

$I_{circuit} = \dfrac{5}{10\,000} = 5 \times 10^{-4}\ \text{A}$

$V_{4k\Omega} = I_{circuit} \times R_{4k\Omega} = 5 \times 10^{-4} \times 4000$
$= 2\ \text{V}$

30 A circuit is set up as shown.

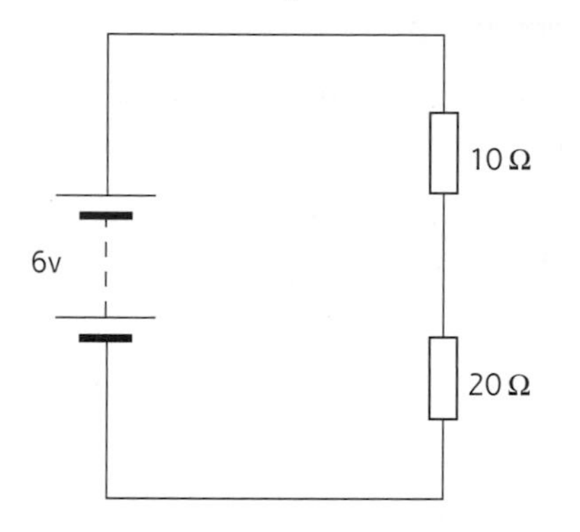

a) Calculate the total resistance of the circuit.

b) Calculate the current in the circuit.

c) Find the voltage across the:
i) 10 Ω resistor
ii) 20 Ω resistor.

31 Circuits are set up as shown.

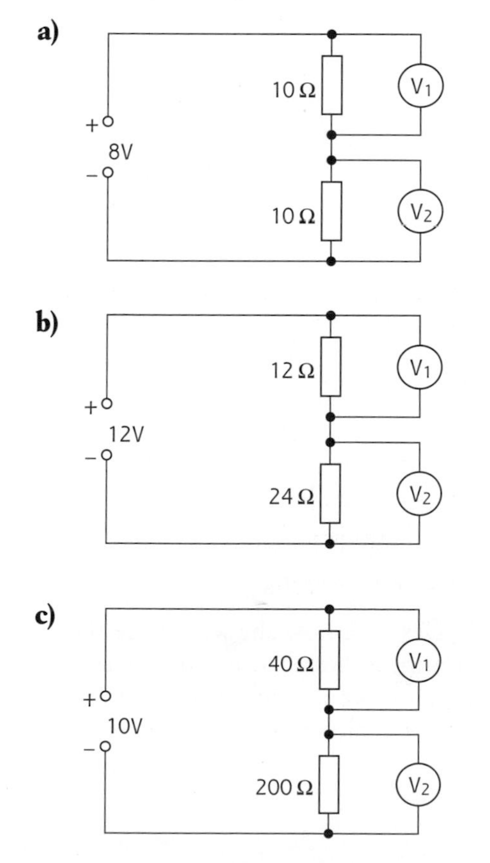

Find the readings on voltmeters V_1 and V_2 in each circuit.

32 Circuits are set up as shown.

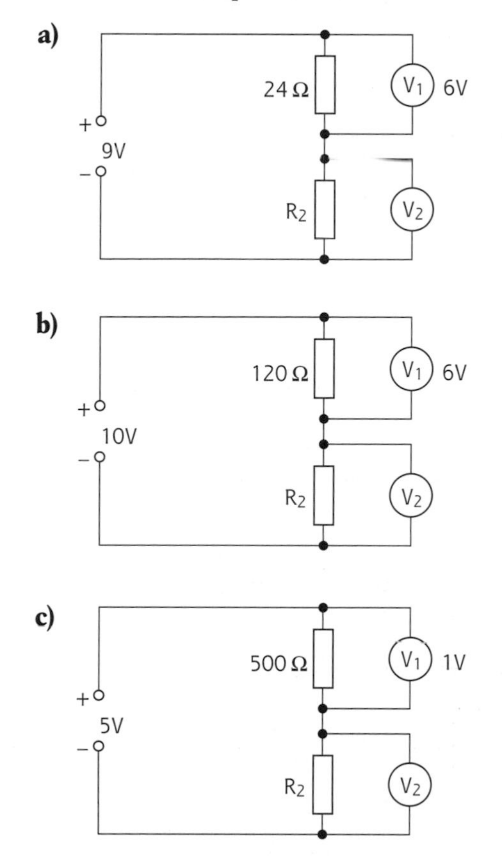

Find the reading on voltmeter V_2 and the resistance of resistor R_2 in each circuit.

33 Circuits are set up as shown.

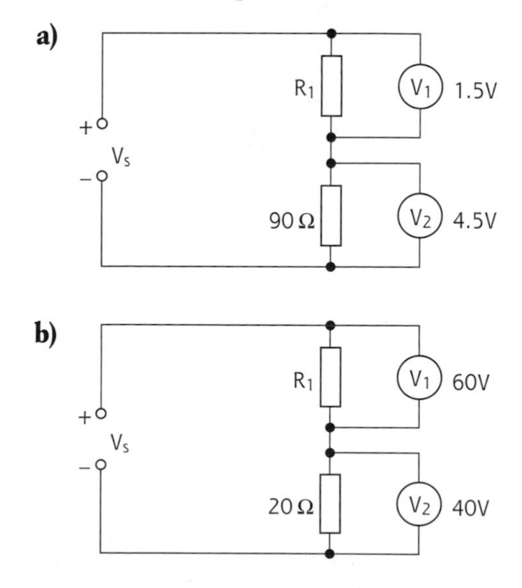

c)

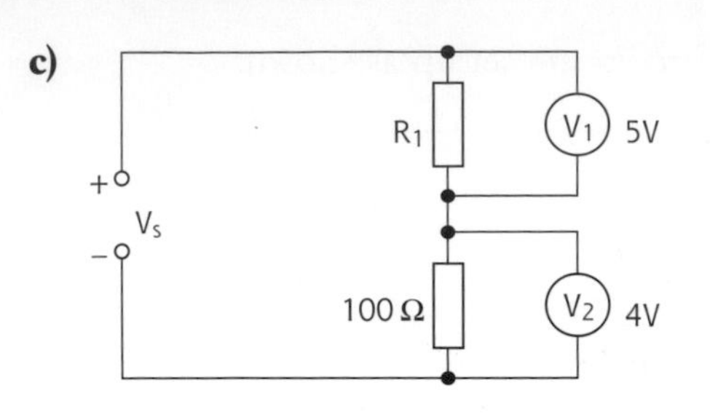

Find the resistance of resistor R_1 and the supply voltage V_S in each circuit.

34 To allow a table lamp to light, two switches must be switched on: the mains switch and lamp switch.

a) How are the switches connected, in series or parallel?

b) Name **two** other household appliances, which require two switches to allow them to operate.

35 The courtesy light of a car has to come on when either the driver's door or the passenger's door or both doors are opened. You may assume that when a door is opened a switch closes.

Draw a circuit diagram that would allow the car courtesy light to work as required.

36 A car has four sidelights. The sidelights can be switched on or off by a single switch. When a lamp 'blows' the other lamps are unaffected.

Draw a circuit diagram to show how the sidelights are connected to the switch and a battery.

37 The two headlights of a car can only be switched on when the headlight switch is on and the ignition switch is on. When a headlight fails the other headlight is unaffected.

Draw a circuit diagram to show how the headlights are connected to the two switches and a battery.

38 The circuit symbols for some electrical components are shown.

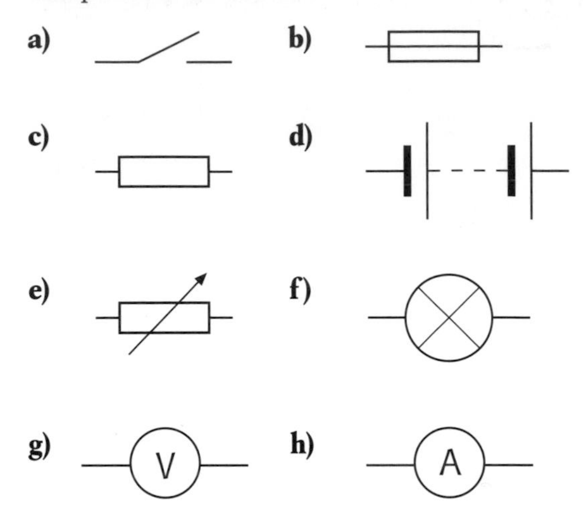

Name each of the components.

39 A girl suspects that the fuse in the 13 A plug attached to her stereo has blown. In order to test the fuse, she decides to make a simple continuity tester as shown.

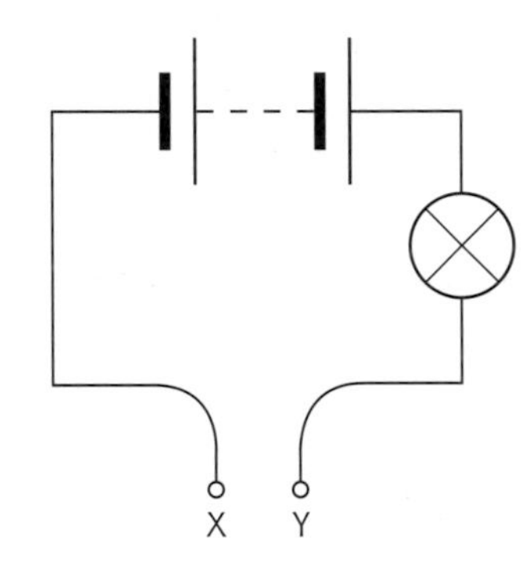

a) How are the components connected, in series or parallel?

b) How would she know that the continuity tester was working?

c) **i)** Describe how she would use the continuity tester to test the fuse.
ii) How would she know whether the fuse was blown or not?

1.2 Electrical energy

- Power = $\frac{\text{energy}}{\text{time}}$ i.e. $P = \frac{E}{t}$
- Power = current × voltage i.e. $P = IV$
- Power = current^2 × resistance i.e. $P = I^2R$
- Power = $\frac{\text{voltage}^2}{\text{resistance}}$ i.e. $P = \frac{V^2}{R}$
- Mains supply has a frequency of 50 Hz.
- The declared value of mains voltage is quoted as 230 V.
- The declared value of an alternating voltage is less than the peak value.

QUESTIONS

1 The names of several electrical appliances are shown in the box below.

food mixer electric fire personal stereo digital clock television

Which appliance can be best matched to each of the following energy changes?

a) electrical energy to sound

b) electrical energy to light

c) electrical energy to heat

d) electrical energy to kinetic energy

e) electrical energy to light and sound.

2 Write down the main energy change for each of the following household appliances.

a) electric cooker

b) radio

c) kettle

d) washing machine

e) hi-fi.

3 In the following sentence the words represented by the letters A and B are missing.

When there is a current in a resistor or a wire, ___A___ energy is changed to ___B___.

Match each letter with the correct word from the box.

electrical heat joules light resistance

4 In the following sentences the words represented by the letters A, B, C, D, E and F are missing.

Energy is measured in ___A___. Power is measured in ___B___. Power is equal to energy ___C___ by time and so can also be measured in joules in one ___D___. Power is also equal to current ___E___ by ___F___.

Match each letter with the correct word from the box.

current divided joules minute multiplied second voltage watts

Example question

A lamp is connected to 230 V mains supply. The current in the lamp is 0.26 A. Calculate the power of the lamp.

Solution

$P = IV = 0.26 \times 230 = 59.8\text{ W}$

5 In the table shown, calculate the value of each missing quantity.

	Power (W)	Current (A)	Potential difference (V)
a)		3.0	12
b)		5.0	230
c)	120		6
d)	2200		230
e)	120	8.0	
f)	4600	20	

6 A car headlamp is operated from a 12 V car battery. When the headlamp is switched on, the current in the headlamp is 5.0 A.

a) Calculate the power rating of the headlamp.

b) How much electrical energy is used by the headlamp in one second?

7 Calculate the power of each of the following appliances, which are connected to a 230 V mains supply.

a) A television, when the current in the television is 0.2 A.

b) An electric kettle, when the current in the element is 10 A.

c) An electric toaster, when the current in the element is 4.0 A.

8 A shop sells 100 W filament lamps.

a) i) Write down the useful energy change for the filament lamp.
ii) State where this energy change occurs.

b) When switched on how much energy does the 100 W filament lamp use in one second?

9 The element of an electric heater has a resistance of 50 Ω. The heater is plugged into the 230 V mains supply and switched on.

a) i) Write down the main energy change which takes place in the heater.
ii) State where this energy change occurs.

b) Calculate the current in the element.

c) Find the power rating of the heater.

10 A lamp is connected to an electrical supply. When the potential difference across the lamp is V volts and the current in it is I amperes, the lamp has a resistance of R ohms. The power rating, P, of the lamp in watts is given by $P = IV$.

a) Show that the power rating of the lamp can also be given by $P = I^2R$.

b) Show that the power rating of the lamp can also be given by $P = \frac{V^2}{R}$.

Example question

A 10 Ω resistor is connected to a 12 V supply. Calculate the power given out by the resistor.

Solution

$$P = \frac{V^2}{R} = \frac{12^2}{10} = \frac{144}{10} = 14.4\,\text{W}$$

11 In the table shown, calculate the value of each missing quantity.

	Power (W)	Voltage (V)	Resistance (Ω)
a)		12	100
b)		230	20
c)	50	6	
d)	1500	230	
e)	48		3
f)	40		2.5

12 The flex of an appliance has a resistance of 0.2 Ω. The maximum safe current in the flex is 6.0 A. Calculate the maximum power rating of the flex.

13 In the following sentences the words or abbreviations represented by the letters A, B, C, D, E and F are missing.

When a battery is connected to an electrical circuit, electrons move in ___A___ direction round the circuit. Since the current in the circuit is in one direction only it is called ___B___ current (___C___).

When the mains supply is connected to an electrical circuit, electrons move in one direction, then in the ___D___ direction and then back again. This to and fro movement of the current is called ___E___ current (___F___).

Match each letter with the correct word or abbreviation from the box.

alternating	a.c.	direct	d.c.
each	one	opposite	same

14 In the following sentences the numbers represented by the letters A and B are missing.

Mains frequency is ___A___ Hz. The declared value of the mains supply is quoted as ___B___ V.

Match each letter with the correct number from the box.

10	50	80	115	230

15 In the following sentences the words represented by the letters A, B and C are missing.

The declared value of an alternating voltage is ___A___ than its peak value. Therefore, the ___B___ value of the mains voltage is greater than its ___C___ value of 230 V.

Match each letter with the correct word from the box.

declared	greater	less	peak

16 Describe the movement of charge in an electrical circuit when the circuit is connected to:

a) a d.c. supply

b) an a.c. supply.

17 An alternating voltage supply has a quoted value of 120 V. Give a possible value for the peak voltage for this supply.

18 A microphone is connected to an amplifier. When the p.d. across the microphone is 3 mV, the microphone supplies 1.2×10^{-10} J of energy every second to the amplifier.

a) What is the input power to the amplifier?

b) Calculate the resistance of the microphone.

19 An amplifier provides a power output of 9.6 W to a loudspeaker. The loudspeaker has a resistance of 15 Ω. Calculate the potential difference across the loudspeaker.

20 Write down the names of three electrical appliances which change electrical energy into heat.

1.3 Electromagnetism

- A magnetic field surrounds a current-carrying wire and is controlled by the current in the wire.
- A voltage (p.d.) can be produced across the ends of a coil when the magnetic field near the coil changes.
- The size of the voltage (p.d.) across the ends of a coil can be increased by increasing the strength of the magnetic field, increasing the number of turns on the coil and increasing the speed of the coil as it moves through the magnetic field.
- A transformer consists of two separate coils of wire wound on an iron core.
- Transformers are used to change the size of an a.c. voltage.
- For a 100 per cent efficient transformer $\frac{V_p}{V_s} = \frac{N_p}{N_s} = \frac{I_s}{I_p}$.
- High voltages are used in the transmission of electrical power as this reduces the current that the transmission lines carry and so the power loss ($= I^2R$) is reduced.
- Electrical power is distributed by the National Grid system – electrical power produced at the power station at 25 000 V is stepped-up by a transformer to 400 000 V for efficient transmission along the transmission lines. At the other end of the transmission lines a step-down transformer reduces the voltage to 230 V for use in our homes.

QUESTIONS

1 Diagram A, below, shows the direction of four compass needles before a current-carrying wire is placed between them. Diagram B show the effect the current-carrying wire has on the compass needles.

A

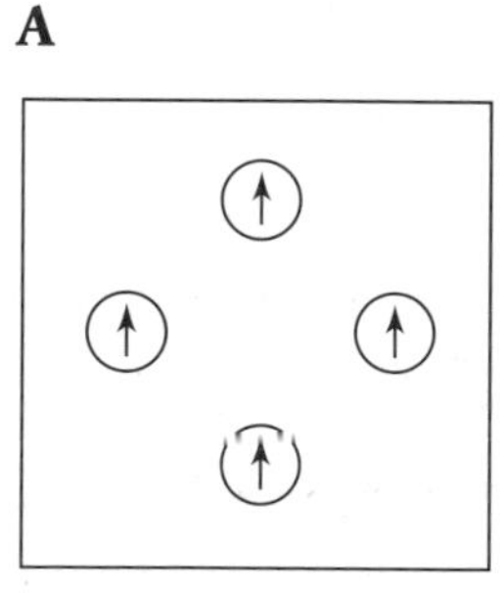

B

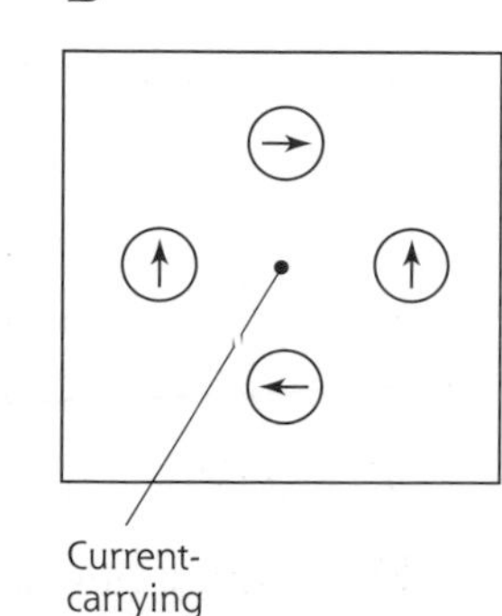

a) Why do the compass needles move when the current-carrying wire is placed between them?

b) What would be the effect on the direction of the compass needles in diagram B if the current in the wire
i) is switched off?
ii) is reversed?

2 A pupil uses a power supply, a length of wire and a nail to produce an electromagnet. She does this by winding a short length of the wire around the nail 25 times. She then attaches the ends of the wire to a 2 V d.c. supply. She finds that the electromagnet is able to pick up five paper clips.

Suggest **two** changes she could make that would allow the electromagnet to lift more paper clips.

3 A circuit is set up as shown.

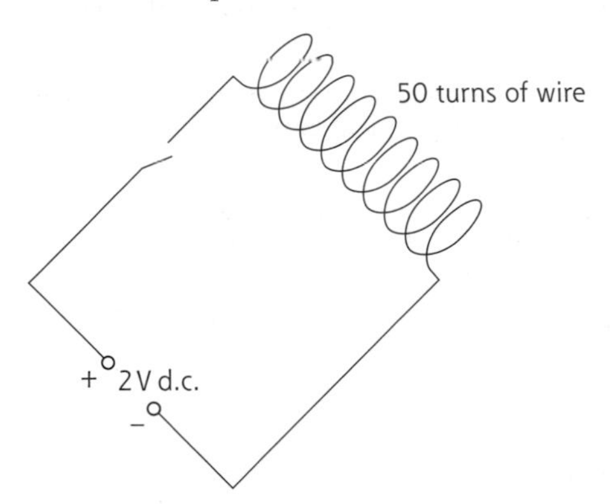

The switch is closed. The 50 turns of wire now form a weak electromagnet.

Suggest **three** changes which would increase the strength of the electromagnet.

4 The diagrams show a loop of wire connected to a voltmeter.

A) **B)**

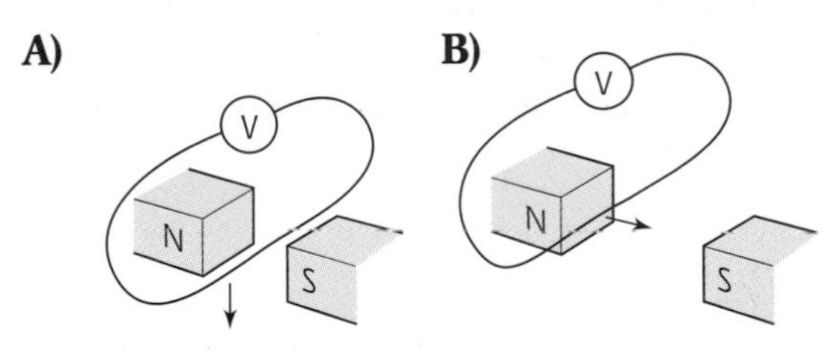

C)

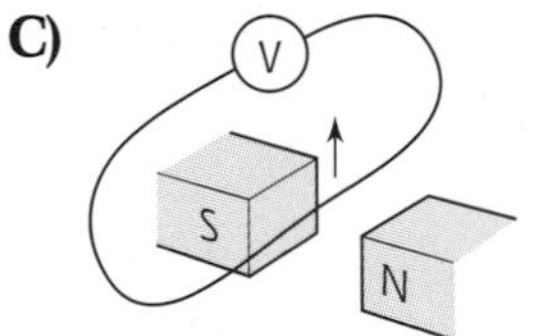

D)

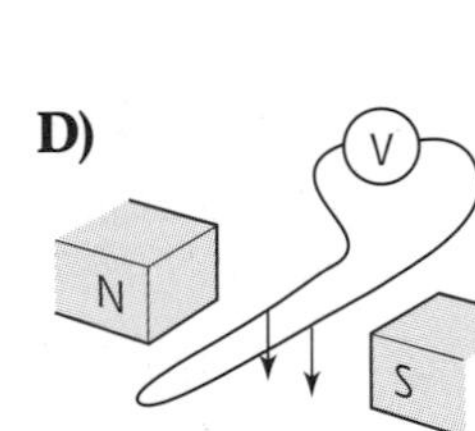

E)

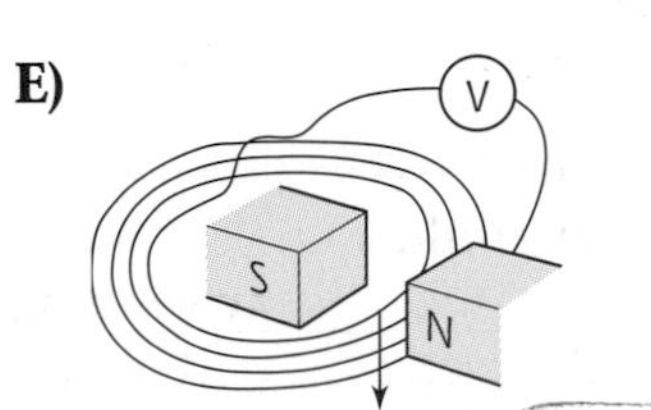

The arrows indicate the direction of movement of the wire between the poles of a magnet.

a) In which diagram(s) will there be a reading on the voltmeter.

b) In diagram E, what changes could be made to increase the reading on the voltmeter?

5 The diagram shows a magnet mounted on an axle.

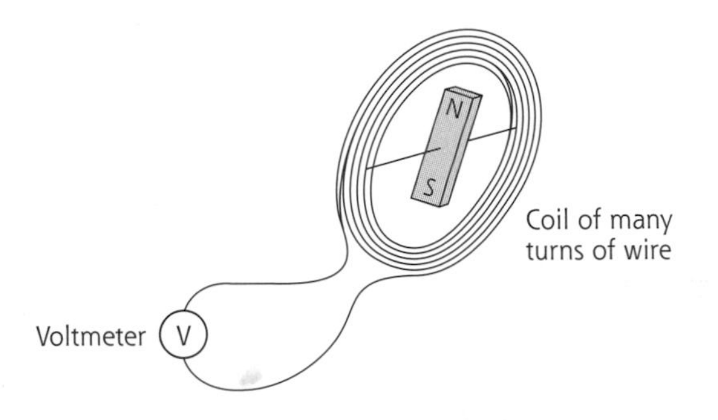

The magnet can be rotated near a coil of wire. The coil is connected to a voltmeter.

Describe what happens to the reading on the voltmeter when the magnet is rotated slowly near the coil.

6 In the following sentences the words represented by the letters A, B, C and D are missing.

A transformer consists of two separate coils (called the primary and secondary coils) wound on an iron core. Changing the current in the primary coil causes a changing ___A___ field around the primary coil. Since the iron core links the two coils this changing magnetic field causes a ___B___ to be produced across the ___C___ coil. Transformers change the size of an ___D___ voltage and transformers only work when connected to an a.c. source.

Match each letter with the correct word from the box.

alternating	**direct**	**magnetic**
primary	**secondary**	**voltage**

7 A student is asked to make a transformer.

a) State the apparatus the student requires.

b) Describe how the student would make the transformer.

Example question

The following information is given for a transformer:

primary coil = 400 turns
secondary coil = 1200 turns
p.d. across primary coil = 10 V.

Calculate the p.d. across the secondary coil assuming the transformer is 100 per cent efficient.

Solution

$$\frac{V_S}{V_P} = \frac{N_S}{N_P}$$

$$\frac{V_S}{10} = \frac{1200}{400}$$

$$V_S = \frac{1200}{400} \times 10 = 30\,\text{V}$$

8 In the table shown, calculate the value of each missing quantity (assume that the transformers are all 100 per cent efficient).

	V_p (V)	V_s (V)	N_p	N_s
a)		23	100 turns	10 turns
b)	10		750 turns	1500 turns
c)	12	120		400 turns
d)	24	6	500 turns	

9 A transformer is connected in each of the three circuits as shown.

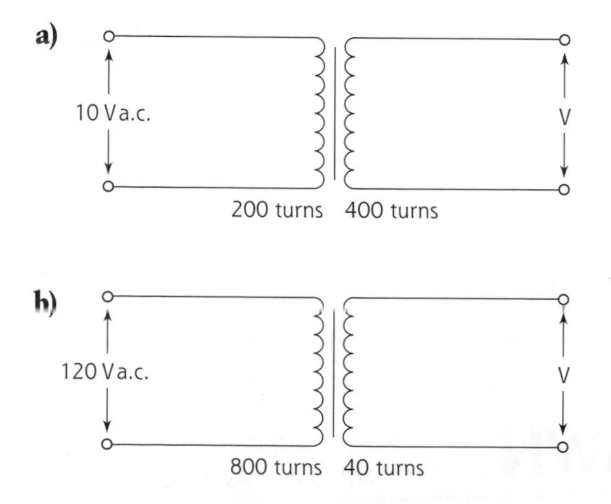

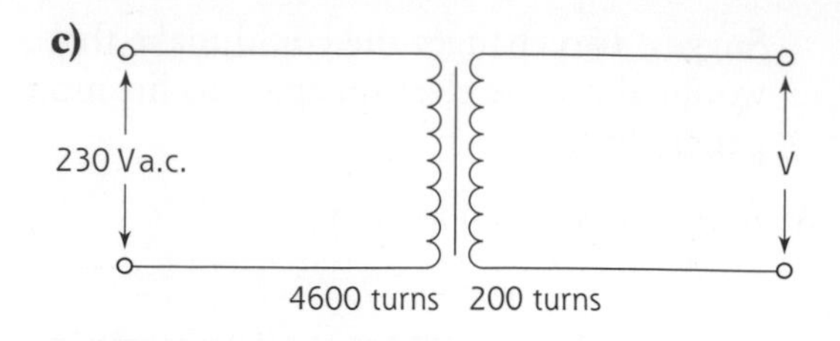

i) Calculate the output voltage from each transformer.

ii) What have you assumed about the transformer used in each circuit?

10 A transformer is connected in each of the three circuits as shown.

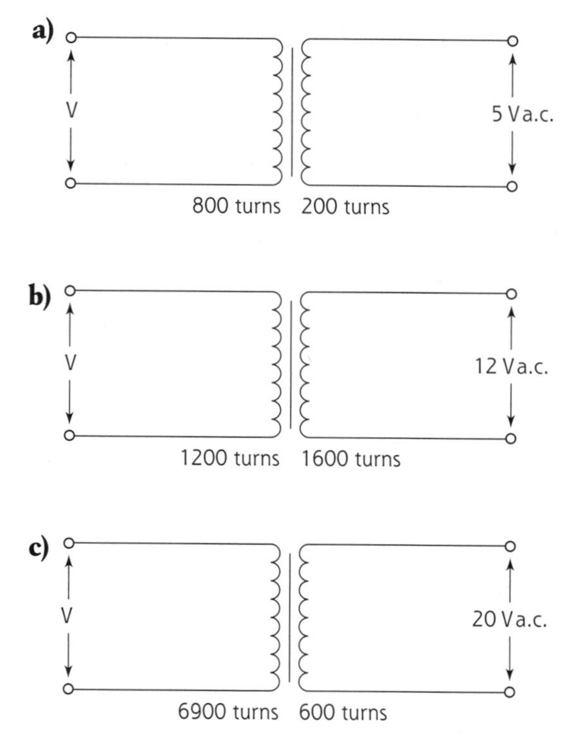

i) Calculate the input voltage for each transformer.

ii) What have you assumed about the transformer used in each circuit?

11 A transformer is connected in each of the three circuits as shown.

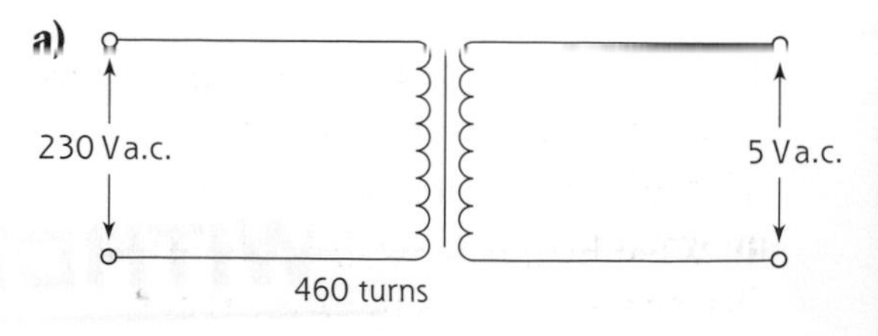

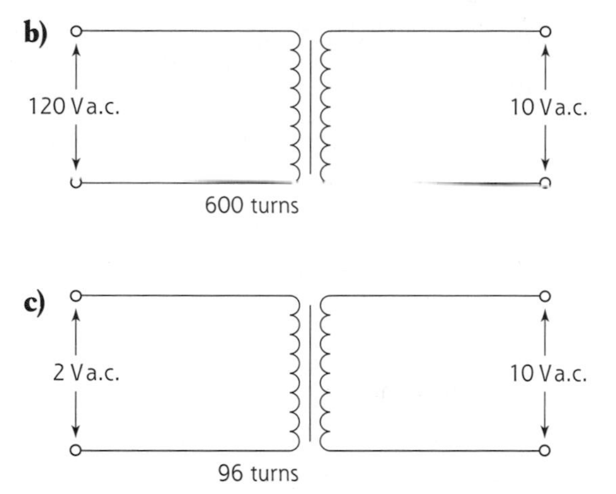

i) Calculate the number of turns that would be required on the secondary coil of each of transformer.

ii) What have you assumed about the transformer used in each circuit?

12 A transformer is connected in each of the three circuits as shown.

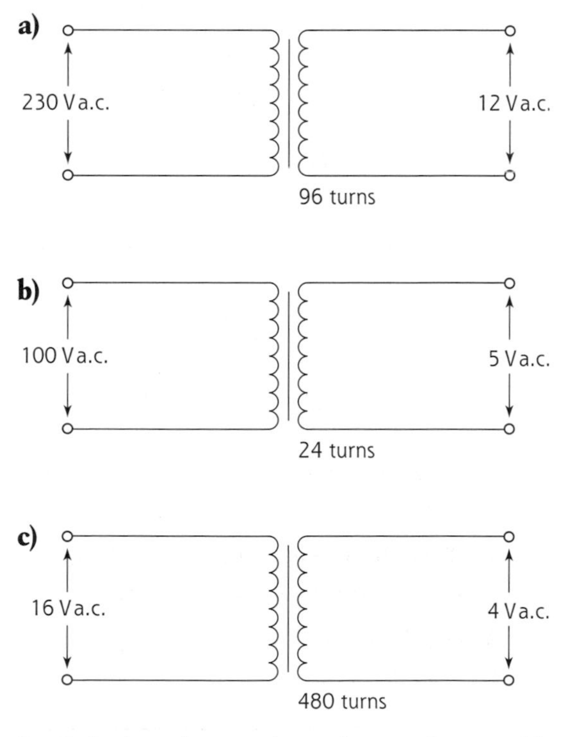

i) Calculate the number of turns that would be required on the primary coil of each transformer.

ii) What have you assumed about the transformer used in each circuit?

Example question

The following information is given for a transformer:

p.d. across primary coil = 15 V
p.d. across secondary coil = 3 V
current in primary coil = 0.3 A.

Calculate the current in the secondary coil assuming the transformer is 100 per cent efficient.

Solution

$$\frac{I_S}{I_P} = \frac{V_P}{V_S}$$

$$\frac{I_S}{0.3} = \frac{15}{3}$$

$$I_S = \frac{15}{3} \times 0.3 = 1.5\text{ A}$$

13 In the table shown, calculate the value of each missing quantity (assume that the transformers are all 100 per cent efficient).

	V_p (V)	V_s (V)	I_p (A)	I_s (A)
a)		12	0.5	2
b)	230		0.02	0.46
c)	100	25		1.2
d)	4000	120	0.03	

14 Calculate the minimum current in the primary coil of the transformers in the circuits shown.

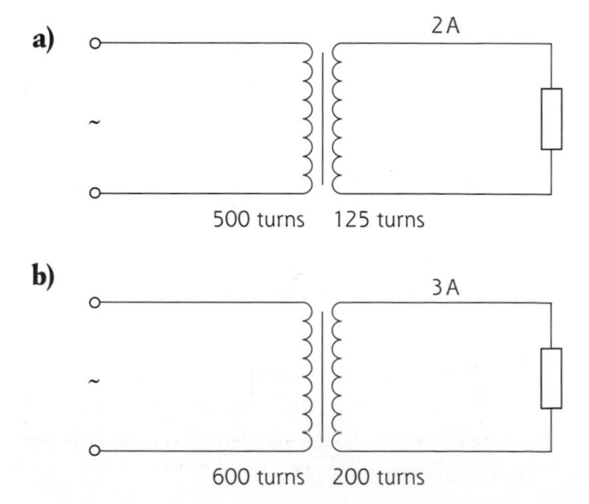

c)

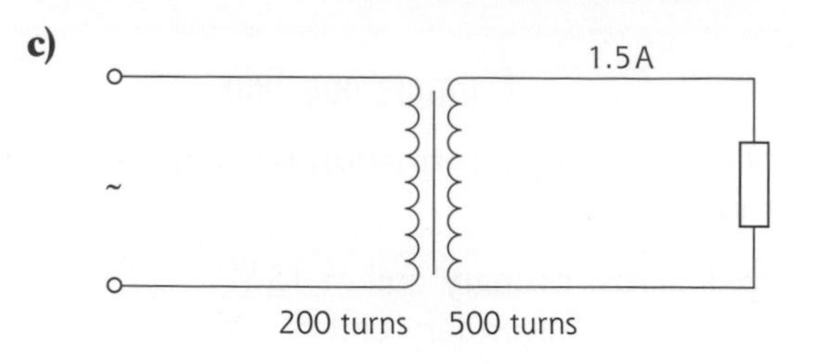

15 Calculate the minimum current in the secondary coil of the transformers in the circuits shown.

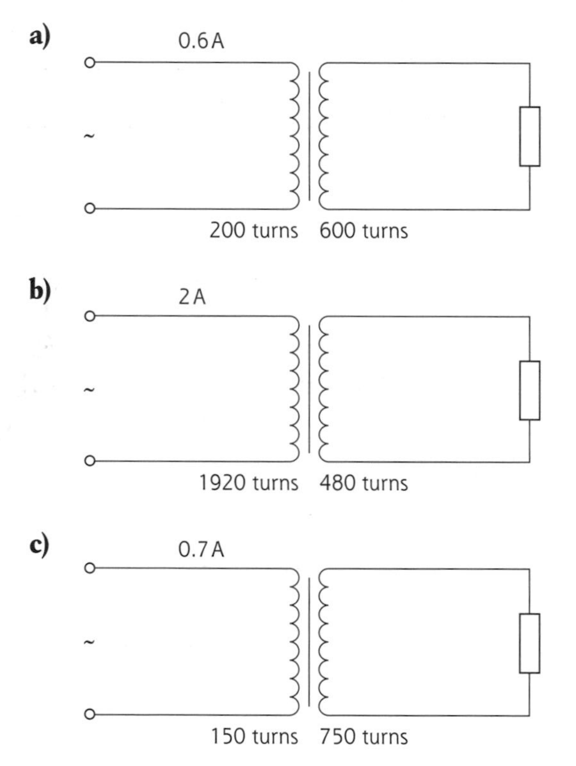

16 A 230 V a.c. supply is connected to the 9200 turn primary coil of a transformer as shown.

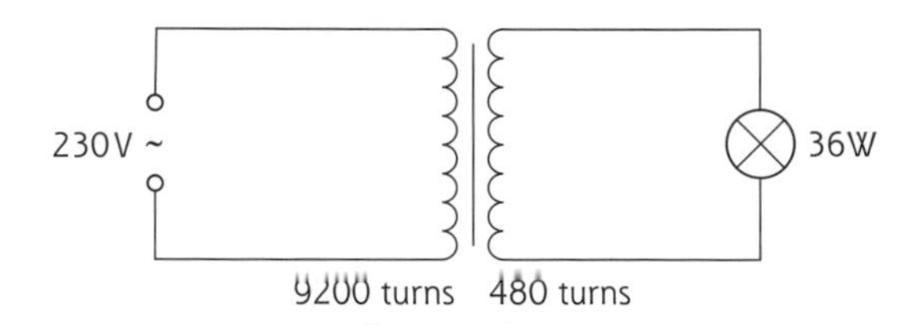

A 36 W lamp is connected to the 480 turn secondary coil of the transformer. The transformer can be assumed to be 100 per cent efficient.

a) Calculate the voltage across the lamp.

b) Calculate the current in the lamp.

c) Find the current in the primary coil of the transformer.

17 The motor to power the pump in a fish tank is designed to work at 12 V and 0.25 A. A circuit containing a transformer is connected to the motor as shown.

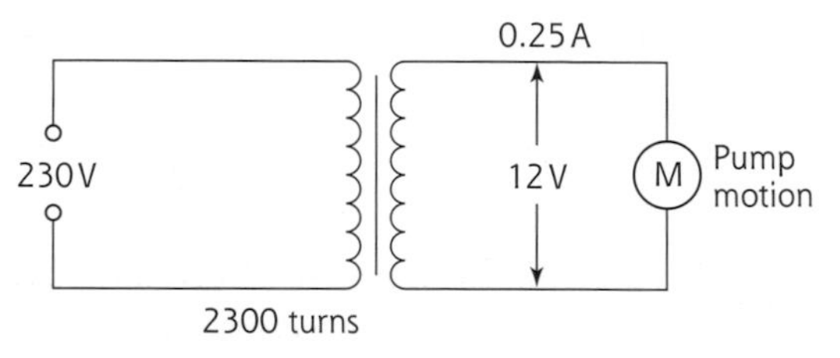

The transformer is 100 per cent efficient.

a) Is the supply a.c. or d.c.?

b) Calculate the number of turns on the secondary coil of the transformer.

c) Calculate the current in the primary coil of the transformer.

18 A diagram of a model transmission system is shown.

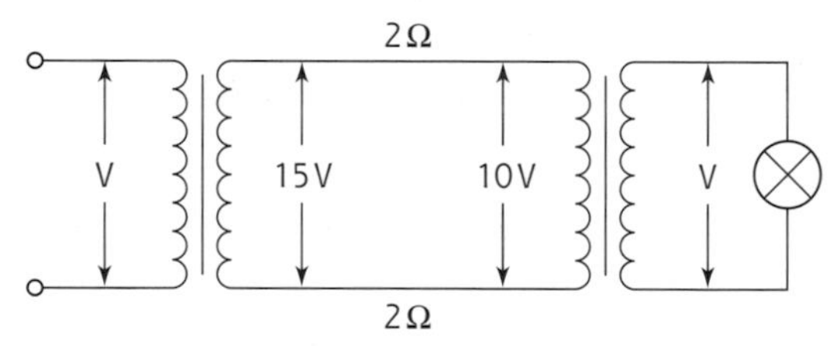

The transformers can be considered to be 100 per cent efficient.

The transmission lines each have a resistance of 2 Ω.

a) Calculate the voltage across the transmission cables.

b) Calculate the current in the transmission cables.

c) Calculate the power loss in the transmission cables.

19 High voltages, such as the 400 000 V used by the National Grid system, are extremely dangerous. Why are such high voltages used in the transmission of electrical energy?

20 Transformers are not 100 per cent efficient. Give **two** reasons why a transformer is not 100 per cent efficient.

21 The National Grid system consists of a number of component parts and different voltages. The diagram illustrates the transmission of electrical energy by the National Grid network.

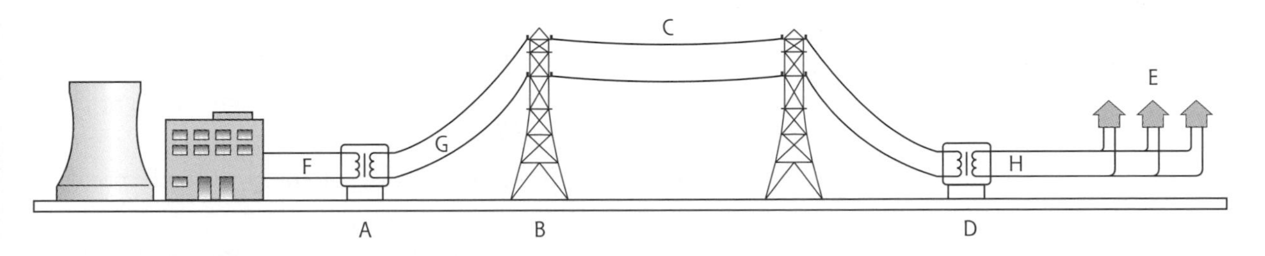

Match letters A to E with the correct component part and match letters F to H with the correct voltage from the box.

homes pylons step-down transformer step-up transformer transmission cables 230 V 25 000 V 400 000 V

22 A wind farm has an electrical power output of 240 MW. This power is transmitted along transmission cables at a voltage of 400 000 V. The transmission cables have a resistance of 0.25 Ω per kilometre.

a) Calculate the current in the transmission cables.

b) The transmission cables have a total length of 100 km. What is the total resistance of the cables?

c) Calculate the power lost in the transmission cables.

d) How much power will be delivered to the destination?

1.4 Electronic components

- An output device changes electrical energy into some other form of energy.
- There are many examples of output devices, such as loudspeaker, relay, filament lamp and light emitting diode (LED).
- LEDs will only light when connected to the power supply the correct way round.
- A resistor should always be connected in series with an LED – this protects the LED from damage from too high a current in it (or too high a voltage across it).
- Most input devices change some form of energy into electrical energy.
- There are a number of input devices, examples are microphone, thermocouple, solar cell, thermistor and light dependent resistor (LDR).
- The resistance of a thermistor changes with temperature – the resistance of most thermistors decreases as the temperature increases.
- The resistance of an LDR decreases with increasing light level.
- An NPN transistor and a MOSFET (metal oxide semiconductor field effect transistor) are electrically operated switches.
- The circuit symbols for an NPN transistor and a MOSFET are shown below:

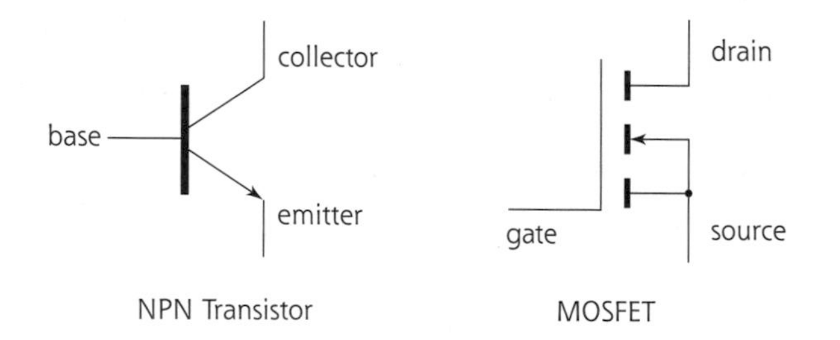

- A transistor is non-conducting (off) for voltages below a certain value but conducting (on) for voltages at or above this certain value.
- An amplifier is a device which makes electrical signals larger.
- Audio amplifiers are found in devices such as radios, televisions, hi-fis, intercoms and loudhailers.
- The output signal from an audio amplifier has the same frequency as, but a larger amplitude than, the input signal to the amplifier.
- For an amplifier: voltage gain $= \dfrac{\text{output voltage}}{\text{input voltage}}$ i.e $V_{gain} = \dfrac{V_o}{V_i}$

QUESTIONS

1 Five output devices are listed below in the box.

buzzer lamp LED motor relay

From the box choose an appropriate output device which could form part of an electrical circuit to:

a) operate a conveyer belt

b) alert a car driver that a door is open

c) indicate that a microwave programme is complete.

2 Five input devices are listed below in the box.

microphone thermocouple
solar cell thermistor
light dependent resistor (LDR)

From the box choose an appropriate input device which could form part of an electrical circuit to be used to:

a) charge batteries during the day

b) adjust the brightness of a television to suit the light level in the room

c) measure the temperature of a Bunsen flame

d) allow a sound to be observed on an oscilloscope.

3 State the energy change that takes place in:

a) a microphone

b) a thermocouple

c) a solar cell.

4 Draw the circuit symbol for a light emitting diode (LED).

5 A girl connects an LED, resistor and a 9 V battery in series. The LED lights.

The girl then makes three changes to the circuit, each change following in turn from the last.

Change X – she reverses the connections to the LED, *then*

Change Y – she reverses the connections to the battery, *then*

Change Z – she connects the LED to the battery but removes the resistor from the circuit.

a) What is the effect on the LED of change X?

b) What is the effect on the LED of changes X and Y?

c) What is the effect on the LED of changes X, Y and Z?

6 A circuit is set up as shown.

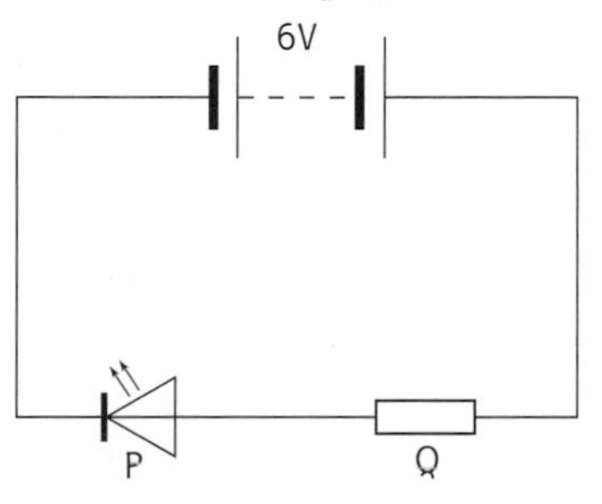

a) Name components P and Q.

b) Explain why component Q is necessary in the circuit.

7 A circuit is set up as shown.

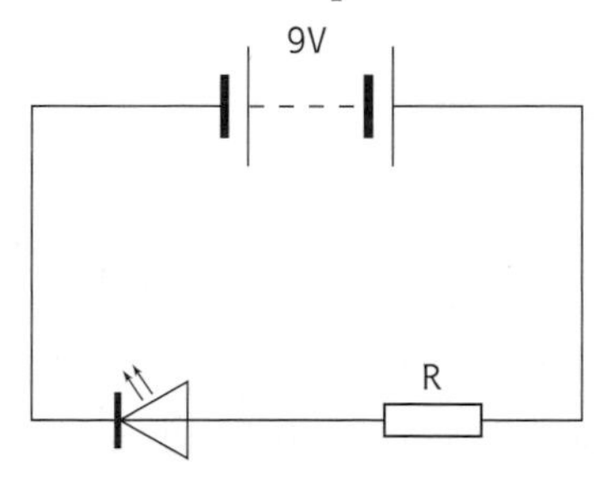

The p.d. across the LED is 1.5 V.

a) How are the components connected, in series or parallel?

b) Calculate the p.d. across the resistor.

c) The current in the circuit is 7.5 mA. Calculate the resistance of the resistor.

8 A number of devices are listed below in the box.

lamp LED loudspeaker
microphone MOSFET solar cell
thermocouple transistor

a) i) State which of the above are input devices.
ii) Give the energy transformation for each input device.

b) i) State which of the above are output devices.
ii) Give the energy transformation for each output device.

9 Name an appropriate input device which could be used in the following:

a) an electronic thermometer

b) a decibel meter (to measure noise level)

c) an electronic light meter.

10 A student is asked to design a circuit to light an LED. The student has the following components: a 6 V battery, a switch, an LED and a resistor.

a) Draw a suitable circuit diagram which will allow the LED to light when the switch is closed.

b) The voltage across the LED must not exceed 1.6 V. The current in the LED must not exceed 8 mA.

Calculate the resistance of the resistor required for this circuit.

11 A student measures the resistance of a thermistor using an ohmmeter. When the thermistor is on the bench the reading on the ohmmeter is 20 000 Ω.

Explain what happens to the reading on the ohmmeter when the student places the thermistor:

a) nearer the window where the light intensity is greater but the temperature is the same

b) close to a warm radiator but where the light intensity is the same.

12 A student measures the resistance of a light dependent resistor (LDR) using an ohmmeter. When the LDR is on the bench the reading on the ohmmeter is 1500 Ω.

The student now moves the LDR to the window where the light intensity is greater. Describe and explain what happens to the reading on the ohmmeter.

13 A student carries out an experiment to find how the resistance of an LDR changes with light intensity. She obtains the results shown in the table.

Light intensity (units)	Voltage across LDR (V)	Current in LDR (A)
250	12	0.005
500	12	0.010
1000	12	0.025

a) Calculate the resistance of the LDR at each of the light intensities.

b) How does the resistance of the LDR depend on light intensity?

c) Draw a circuit diagram of the apparatus that the student could have used to obtain these results.

14 A student uses the apparatus shown to measure the resistance of a thermistor.

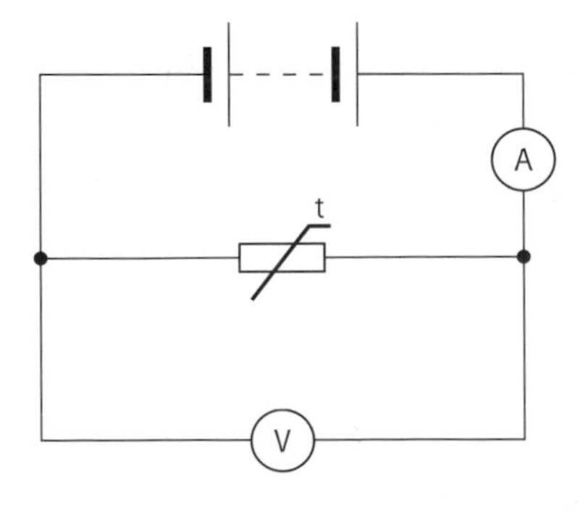

When the temperature of the thermistor is 18 °C the reading on the voltmeter is 10 V and the reading on the ammeter is 0.25 A.

a) Calculate the resistance of the thermistor at this temperature.

b) The resistance of this thermistor decreases as temperature increases. Suggest suitable readings on the ammeter and voltmeter when the thermistor is at a temperature of 15 °C.

15 In the following sentences the words represented by the letters A, B, C and D are missing.

An NPN transistor and a MOSFET are both types of electronic ___A___. They are controlled by a ___B___ or p.d. They are off (___C___) when the voltage is below a certain value and on (___D___) when the voltage is equal to or above the certain value.

Match each letter with the correct word from the box.

conducting current
non-conducting switch voltage

16 An electronic circuit is set up as shown.

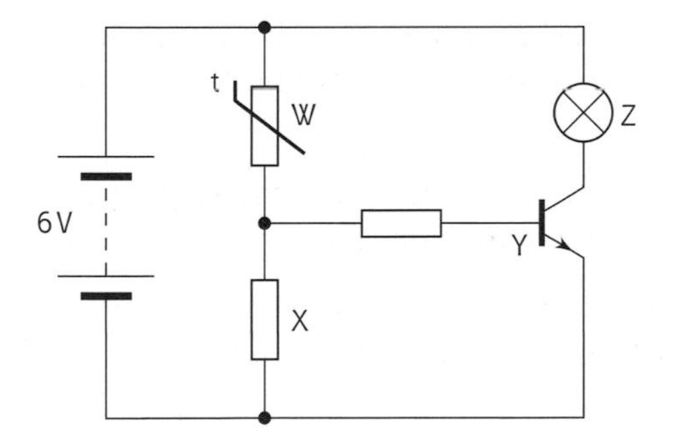

a) Name the components W, X, Y and Z in the circuit.

b) What is the purpose of component Y in the circuit?

17 An electronic circuit is set up as shown.

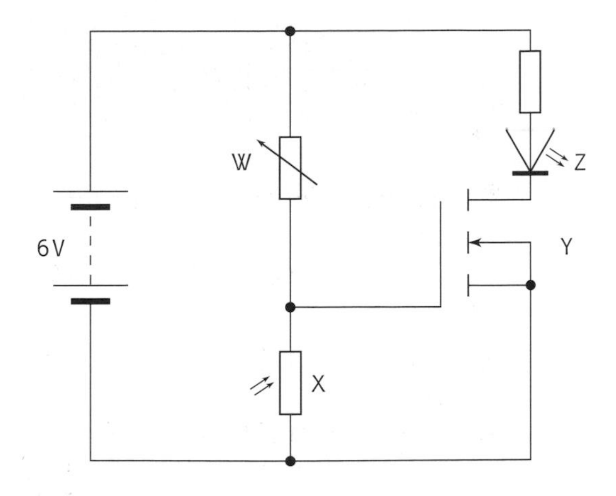

a) Name the components W, X, Y and Z in the circuit.

b) What is the purpose of component Y in the circuit?

18 A student sets up the electronic circuit shown.

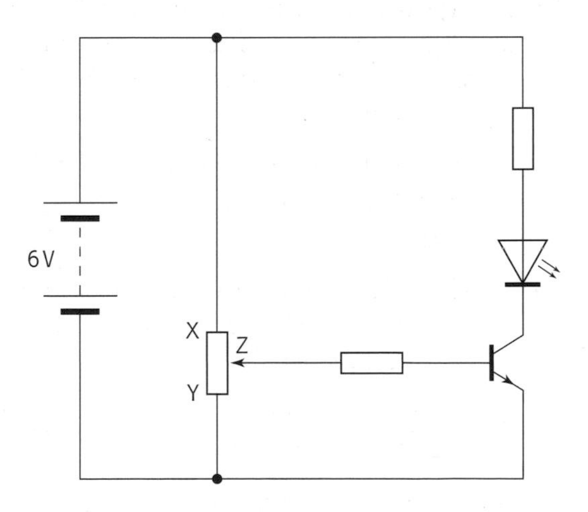

The LED is lit when the control on the potentiometer (Z) is at position X.

a) When the LED is lit, is the transistor on (conducting) or off (non-conducting)?

b) The control on the potentiometer is turned slowly from X to Y. Describe and explain what happens to the LED.

19 A transistor switching circuit is set up as shown.

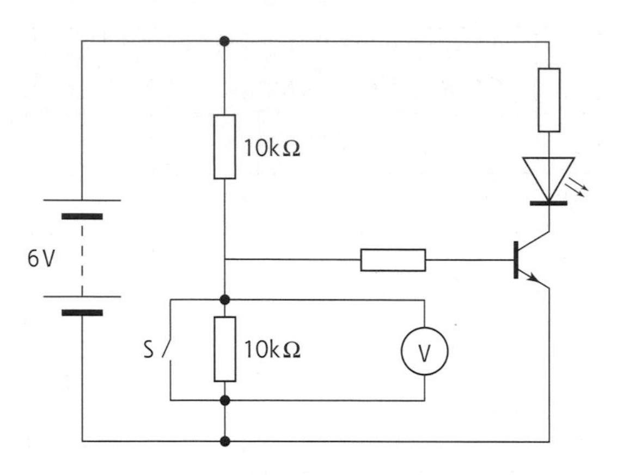

Switch S is closed.

a) **i)** State the reading on the voltmeter.
ii) Is the transistor on (conducting) or off (non-conducting)?
iii) Is the LED lit or unlit?

b) Switch S is now opened. Describe and explain what happens to the LED.

20 Circuits A and B are set up as shown.

A

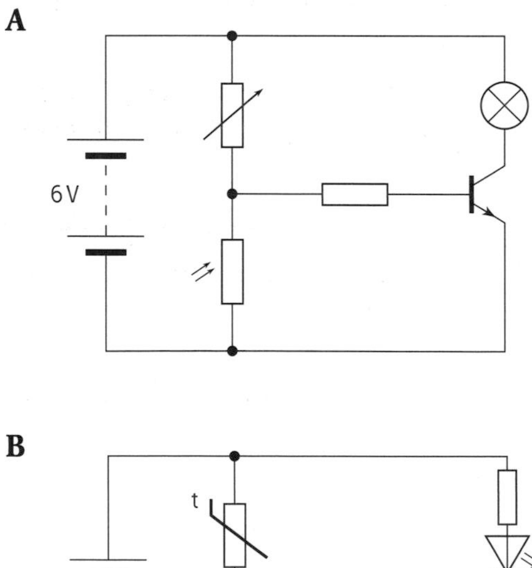

B

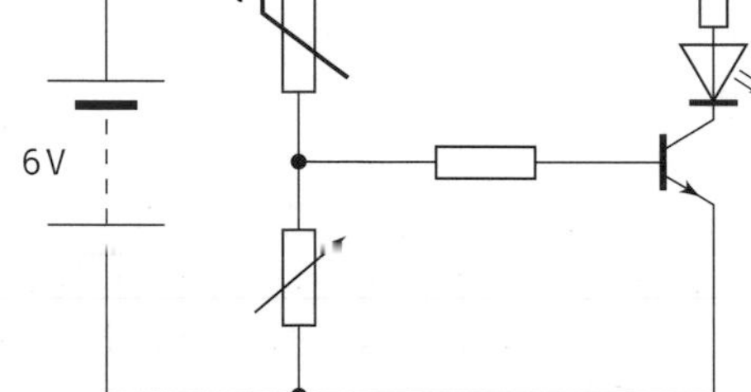

In circuit B the resistance of the thermistor decreases as its temperature increases.

a) Suggest one practical use for each of these circuits.

b) Why is a variable resistor used in each of these circuits instead of a fixed resistor?

21 A student sets up the circuit shown to monitor the temperature in a greenhouse.

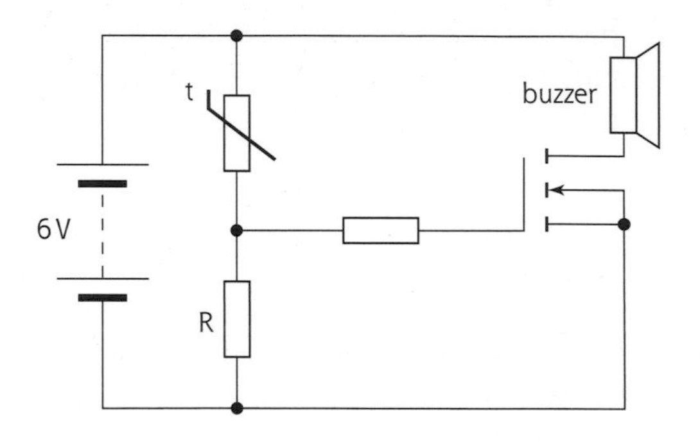

The resistance of the thermistor decreases as its temperature increases.

a) Describe and explain what will happen as the temperature in the greenhouse increases.

b) The student now uses the same components to build a circuit which will sound the buzzer when the greenhouse gets too cold.

What alteration does the student make to the circuit that will allow this to happen?

22 The circuit shown can be used as part of the light sensor for a camera.

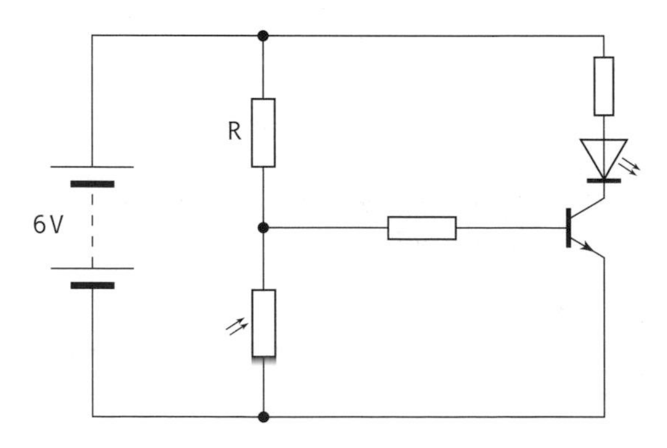

Describe and explain what happens when the camera goes from a brightly lit room to a dimly lit room.

23 Draw the circuit symbol for:

a) an NPN transistor

b) a MOSFET.

24 A number of electrical devices are listed below in the box:

baby alarm electric cooker hi-fi microwave radio television vacuum cleaner washing machine

From the box write down the names of the devices in which amplifiers play an important part.

25 An amplifier is a major component in a radio.

a) What is the purpose of the amplifier in the radio?

b) How does the frequency of the input signal to the amplifier compare with the frequency of the output signal from the amplifier?

c) How does the amplitude of the input signal to the amplifier compare with the amplitude of the output signal from the amplifier?

Example question

The input voltage to an amplifier is 1.5 V. The output voltage from the amplifier is 30 V. Calculate the voltage gain of the amplifier.

Solution

$$V_{gain} = \frac{V_o}{V_i} = \frac{30}{1.5} = 20$$

26 In the table shown, calculate the value of each missing quantity.

	Input voltage (V)	Output voltage (V)	Voltage gain
a)	0.02	8.00	
b)	4×10^{-3}	10	
c)	0.8		30
d)	5.5×10^{-4}		4.2×10^{4}
e)		2.0	10
f)		6.0	5×10^{3}

27 The input voltage to an amplifier is 0.10 V. The output voltage from the amplifier is 20.0 V.

Calculate the voltage gain of this amplifier.

28 The voltage gain of an amplifier is 250. The input voltage to the amplifier is 6.0×10^{-3} V.

Calculate the output voltage from the amplifier.

29 The output voltage from an amplifier is 4.0 V. The voltage gain of the amplifier is 15 000.

Calculate the input voltage to the amplifier.

30 A student is asked to measure the voltage gain of an amplifier.

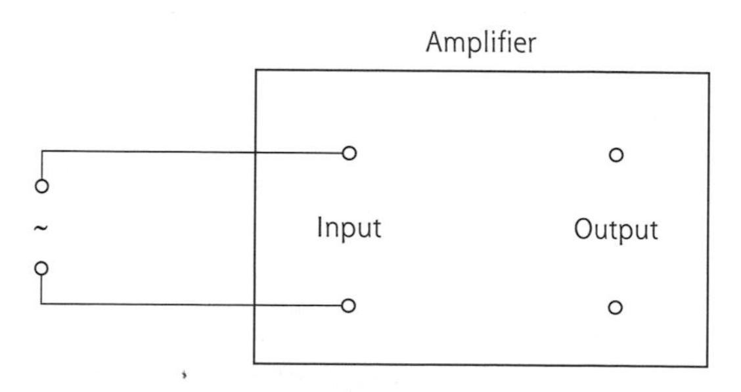

Describe, with the aid of a circuit diagram, how the student measures the voltage gain of the amplifier shown.

Exam-style questions

1 A circuit is set up as shown.

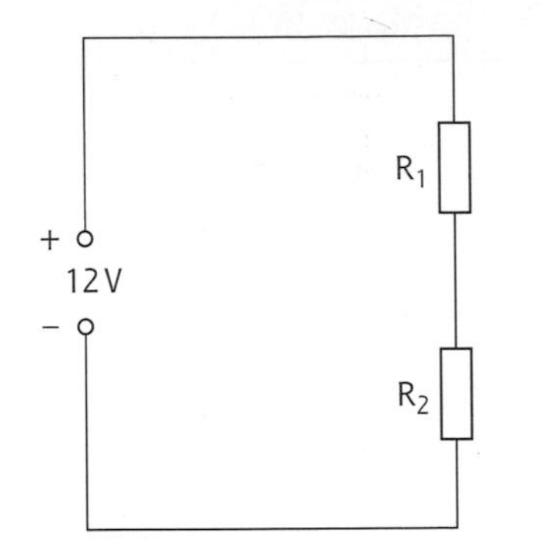

a) What name is given to the arrangement of the two series resistors R_1 and R_2 connected across the 12 V supply?

b) A student is asked to find the resistance of resistor R_2 using an ammeter and a voltmeter.

Redraw the diagram to show how the student should connect the ammeter and the voltmeter in the circuit.

c) The student obtains the following readings:

Ammeter reading = 0.25 A

Voltmeter reading = 3.0 V

Show that the resistance of R_2 is 12 Ω.

d) Calculate the resistance of resistor R_1.

2 Three resistors are connected as shown.

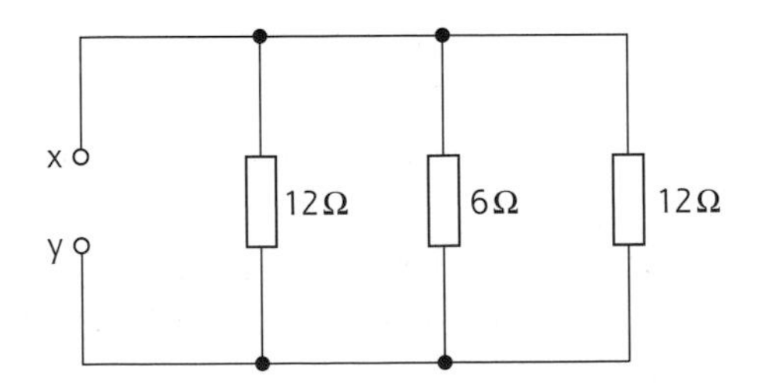

a) Are the resistors connected in series or parallel?

b) Calculate the resistance between X and Y.

c) A 12 V supply is now connected between X and Y. Calculate the current from the supply.

3 a) Look at the following list of electronic devices.

amplifier loudspeaker thermistor
light emitting diode (LED)
light dependent resistor (LDR)
microphone MOSFET
switch transistor

i) From the list, name two output devices.
ii) Draw the symbol for a light emitting diode (LED).
iii) State the main energy transformation which occurs with a microphone.

b) An electronic circuit is set up as shown.

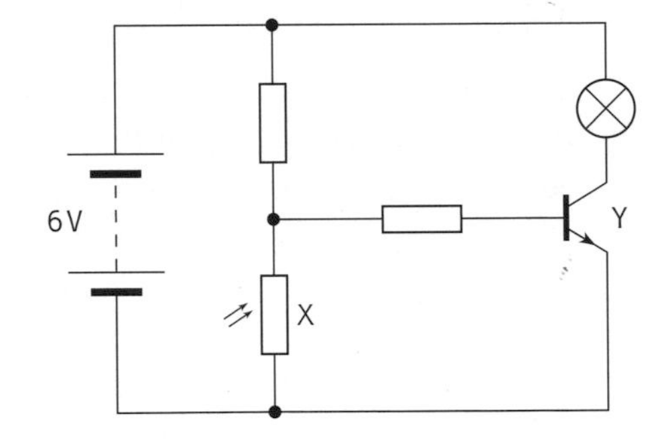

The circuit will switch on a lamp when the light intensity in a room falls below a certain level.
i) Name components X and Y.
ii) Explain how the circuit operates so that the lamp lights when the room becomes too dark.

4 a) A circuit is set up as shown.

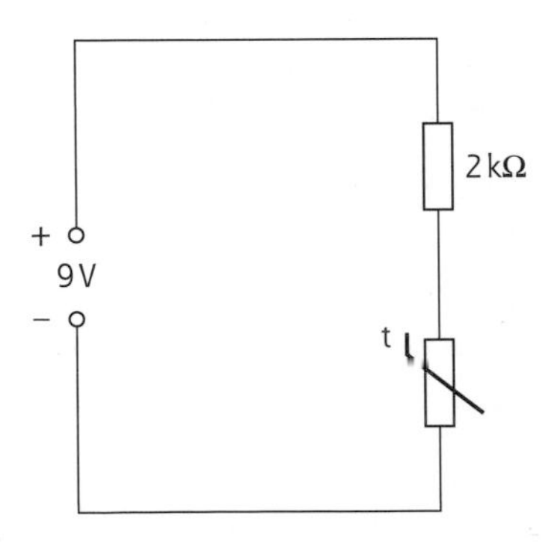

At a certain time during the day, the voltage across the resistor is 6 V.

i) Calculate the current in the resistor.
ii) Show that the resistance of the thermistor is 1000 Ω at this time.
iii) The graph shows how the resistance of the thermistor varies with temperature.

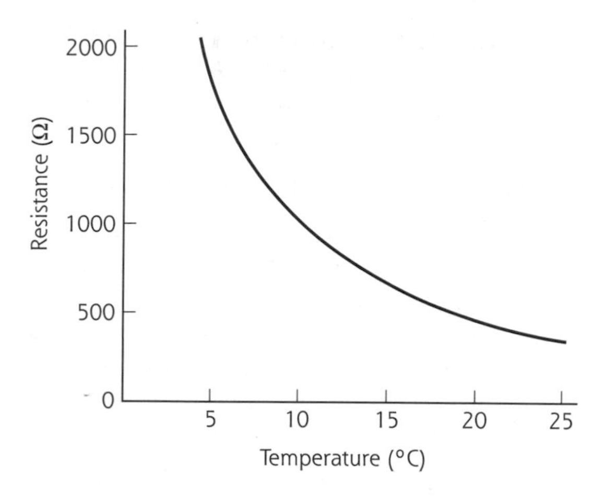

Use the graph to estimate the temperature of the thermistor at this time.

b) The circuit shown below is fitted in a car to alert the driver that the outside temperature is below a certain value.

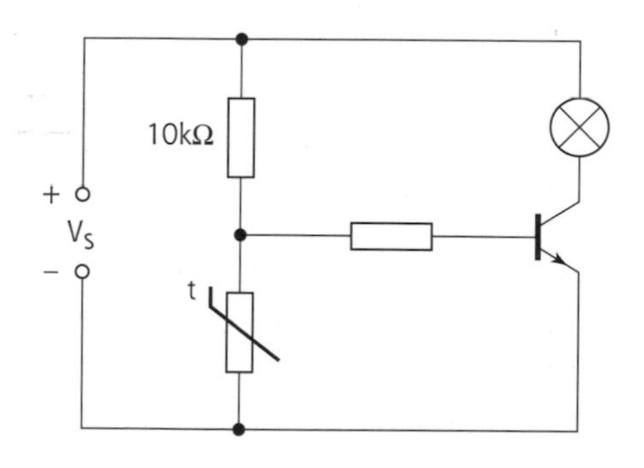

The circuit is designed to switch on a warning lamp when the outside temperature falls below 2 °C.

Describe and explain how the circuit operates so that the warning lamp comes on when the outside temperature falls below 2 °C.

5 The diagram shows part of the electrical wiring for the rear lights of a car.

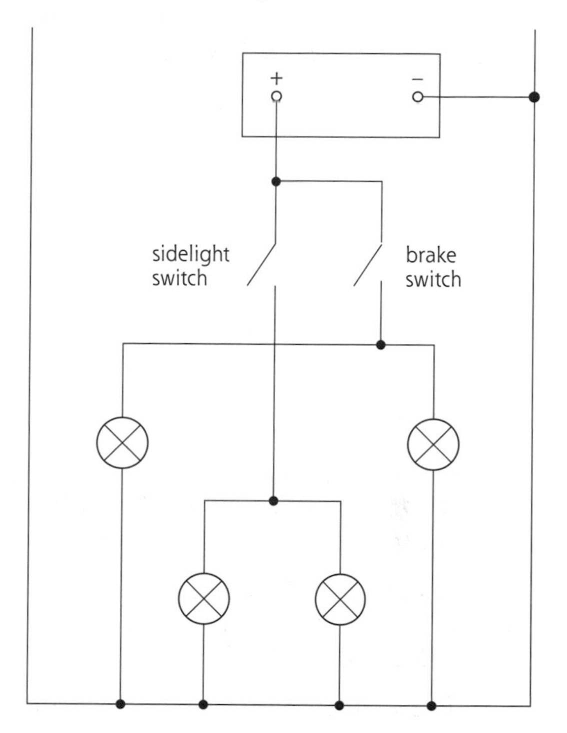

Each rear light contains two lamps. The side-lights are switched on when the front lights are switched on. The brake lights are switched on when the car brakes.

A rear side-light lamp is rated at 12 V, 6 W. A brake lamp is rated at 12 V, 24 W.

a) Explain why the two sets of lamps are connected in parallel rather than in series.

b) Calculate the resistance of:
i) a rear side-light lamp
ii) a brake lamp.

c) How much current is drawn from the battery when:
i) both brake lights only are on?
ii) both rear side-lights and both brake lights are on?

6 Some types of portable electrical equipment can be damaged if the connections from the battery terminals are the wrong way round. The circuit shown indicates whether the terminals of a 9.0 V battery are correctly connected to P and Q.

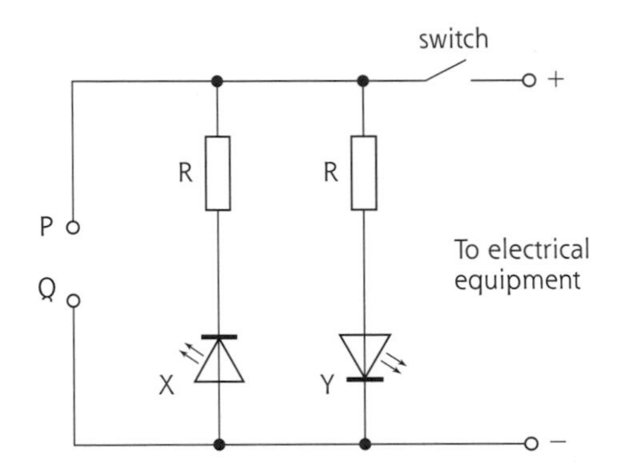

The LEDs are identical. When lit the potential difference across each LED is 1.7 V.

When lit the current in each LED is 10 mA.

a) Which LED, X or Y, would light when the terminals of the 9.0 V battery are connected the correct way round?

b) Calculate the value of resistor, R, which would allow each LED to operate normally.

7 A pupil is asked to measure the voltage gain of an amplifier. The pupil sets up the circuit shown.

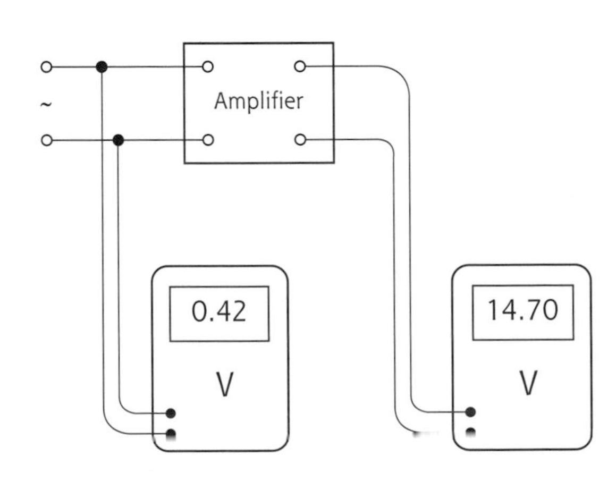

a) Calculate the voltage gain of the amplifier.

b) The input signal to the amplifier has a frequency of 500 Hz. What is the frequency of the output signal from the amplifier?

8 The rating plate of an electric iron is shown.

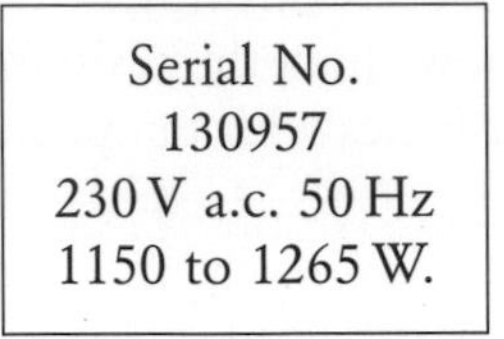

a) What do the initials a.c. stand for?

b) Explain what a.c. means in terms of electron flow in a circuit.

c) Show that the maximum current in the element of the iron is 5.5 A.

d) The iron is switched on for five minutes. Calculate the maximum charge transferred in this time.

9 a) A bar magnet is suspended from a spring. The magnet can oscillate freely in and out of a coil as shown.

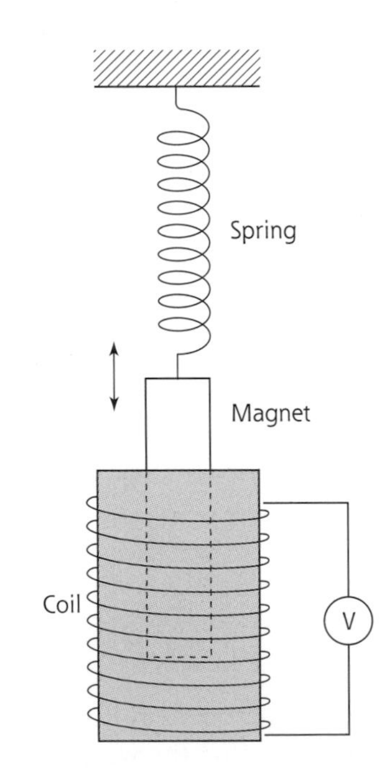

The coil is attached to a voltmeter.

i) The magnet is set oscillating. Describe what happens to the reading on the voltmeter.

ii) Suggest **two** ways in which the size of the reading on the voltmeter could be made larger.

b) Two coils of insulated wire are wound on an iron core to make a transformer as shown.

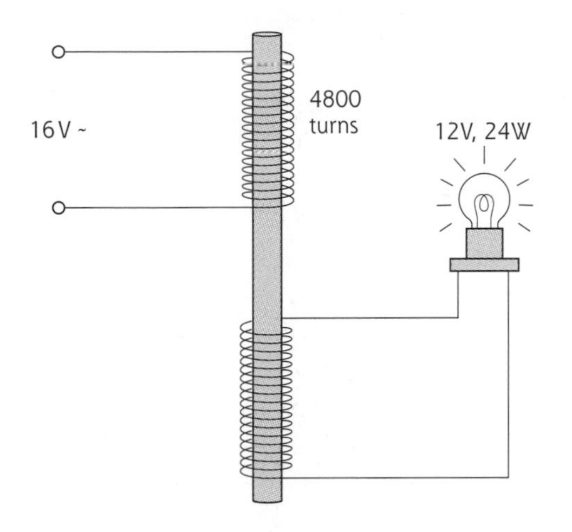

One coil with 4800 turns is connected to a 16 V a.c. supply. The other coil is connected to a lamp. The lamp is operating at its correct rating of 24 W, 12 V.

i) State the purpose of a transformer.

ii) A Calculate the number of turns on the secondary coil.

B State any assumption that you have made.

iii) The current drawn from the 16 V supply is 1.7 A. Calculate the electrical energy lost by the transformer each second.

iv) Suggest **one** reason why a transformer is not 100 per cent efficient.

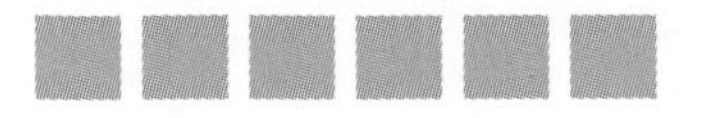

Chapter 2

MECHANICS AND HEAT

2.1 Kinematics

- average speed $= \dfrac{\text{distance}}{\text{time}}$

  Unit of speed is m/s.
- Instantaneous speed is the same calculation as average speed but the time interval is very short. This means that any change in speed is very small.
- Vectors have magnitude and direction but scalars have only magnitude.
- Displacement and velocity are vectors since they have direction as well as magnitude but distance and speed are scalars.
- Acceleration is the change in velocity of an object in one second e.g. an acceleration of $2\,\text{m/s}^2$ means that the object will change its velocity by 2 m/s every second.
- Change in velocity is measured in metres per second (m/s), time for change in velocity to occur is measured in seconds (s) and acceleration is measured in metres per second squared (m/s^2).
- Acceleration $a = \dfrac{v - u}{t}$ where u is initial velocity, v is the final velocity and t is the time for the change in velocity.
- Velocity–time graphs can show:

  a) constant velocity

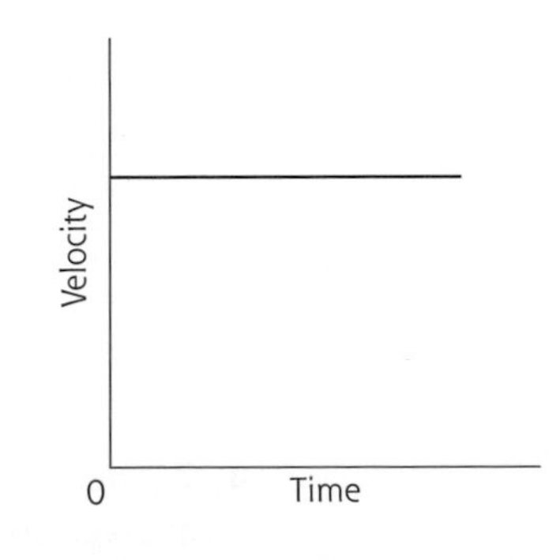

b) constant acceleration

c) constant negative acceleration.

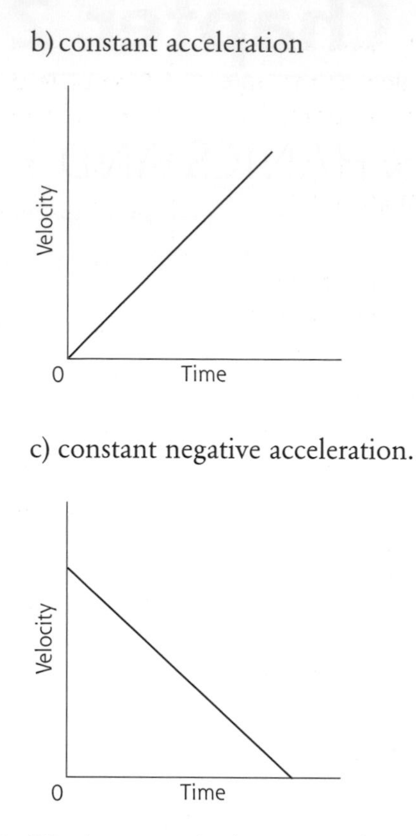

- Displacement is the area under a velocity–time graph.
- Acceleration is the gradient of a velocity–time graph.

QUESTIONS

Example question

An aircraft travels a distance of 642 km in 50 minutes. What is its average speed in m/s?

Solution

$$v = \frac{d}{t} = \frac{642 \times 1000}{50 \times 60} = 214\ \text{m/s}$$

1 You are about to make a journey in a car. You wish to measure the average speed of your journey.

Describe how you would measure the average speed of your journey.

2 In the table shown, calculate the value of each missing quantity.

Distance (m)	Time (s)	Average speed (m/s)
2 000	50	
40 000	75	
	3 600	2 500
	300	0.05
75 000		375
100		12.5

3 A girl on a bicycle travels a distance of 500 m in 125 s. What is her average speed?

4 Between two bus stops, a bus travels at an average speed of 10 m/s. The bus takes 150 s to travel between the stops. How far apart are the bus stops?

5 During a 100 m sprint, a boy's average speed is calculated to be 8.0 m/s. Calculate the time taken by the boy to complete the sprint.

6 A lorry travels a distance of 15 km in 10 minutes. Calculate the average speed of the lorry in metres per second.

7 A train covers a distance of 150 km at an average speed of 40 m/s. Calculate the time taken by the train to cover this distance.

8 A snail travelling at an average speed of 2 mm/s, takes half a minute to cross a leaf. Calculate the distance covered by the snail in this time.

9 Describe how you could measure the instantaneous speed of a bicycle. You are given a tape measure and a stopwatch.

10 **a)** What is the difference between average speed and instantaneous speed?

b) Give two examples of situations where average and instantaneous speeds are different.

Example question

A helicopter travels north for 35 km and then east for 45 km.

Calculate:

a) the distance travelled by the helicopter

b) the displacement of the helicopter.

Solution

Choose a suitable scale and draw two lines with arrows to represent the two vectors. Remember that vectors must add head to tail.

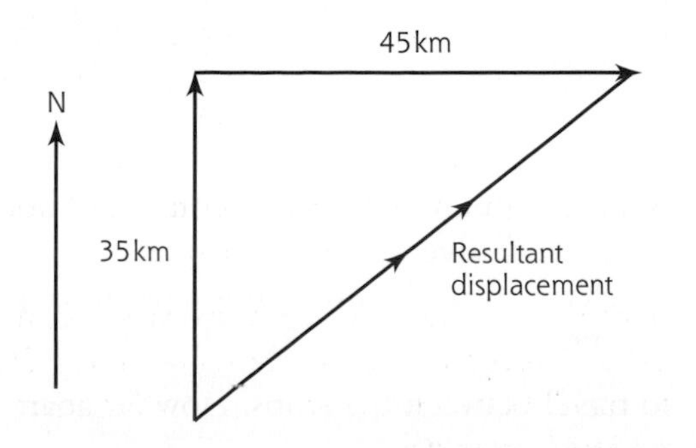

Complete the diagram to show the resultant vector. Measure the line and the angle.

a) distance travelled is 35 + 45 = 80 km.

b) displacement = 57 km (at 52° E of N).

11 **a)** What is meant by a vector quantity?

b) Give one example of a vector quantity.

12 A car travels 3 km east and then 7 km north. Calculate:

a) the distance travelled by the car

b) the displacement of the car.

13 The diagrams show the displacements of a number of different objects.

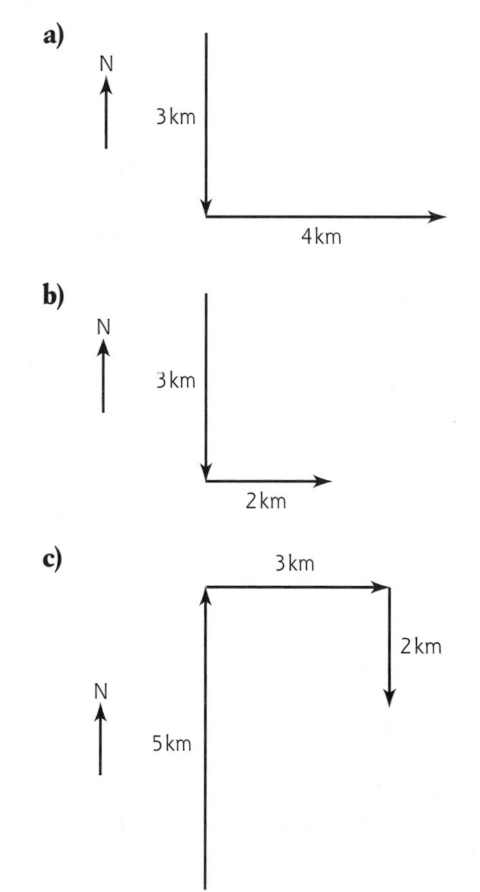

d)

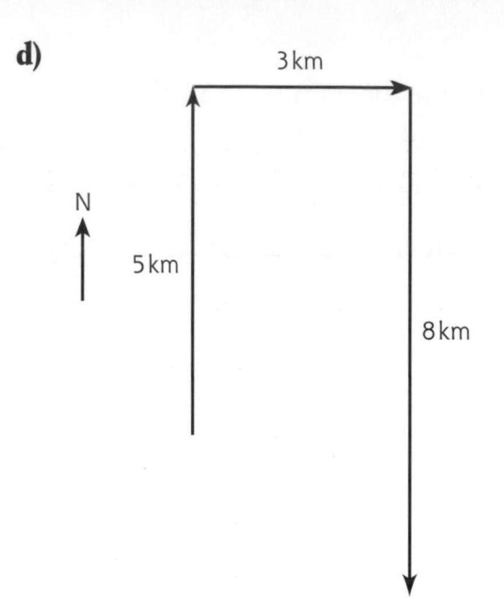

Calculate the resultant displacement in each case.

14 An aircraft flies north at a speed of 120 m/s. A cross wind is blowing east at a speed of 18 m/s. By drawing or calculation, find the resultant velocity of the aircraft.

15 A yacht is cruising due south at a speed of 2.5 m/s. A wind is blowing west at a speed of 0.8 m/s. What is the resultant velocity of the yacht?

16 A rower tries to row straight across a river at 12 m/s in the diagram below. The river is flowing downstream at 4 m/s. Calculate the actual velocity of the rower.

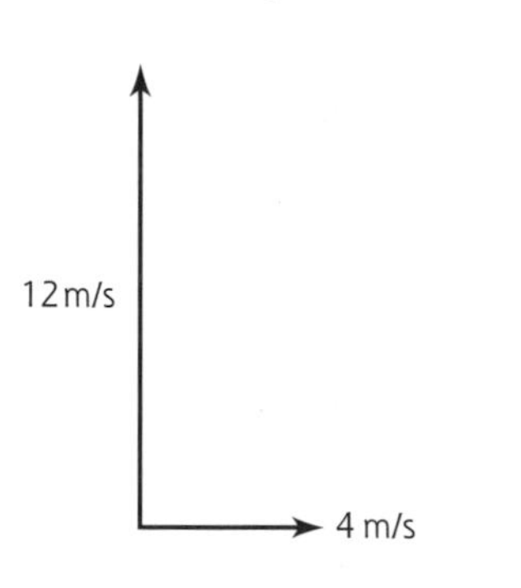

17 A boat travels east in a river at a speed of 4 m/s. The river flows south at a speed of 5.5 m/s. Calculate the resultant velocity of the boat.

Example question

A car is travelling along a straight road at 12 m/s. The car accelerates uniformly for 8.0 s and it reaches a speed of 17 m/s. Calculate the acceleration of the car.

Solution

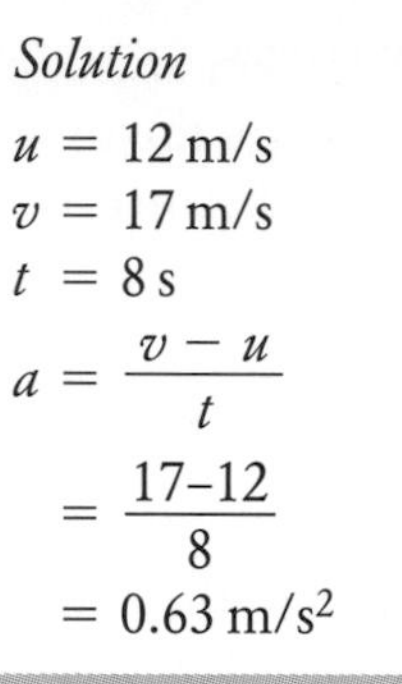

$u = 12\text{ m/s}$
$v = 17\text{ m/s}$
$t = 8\text{ s}$

$$a = \frac{v - u}{t}$$

$$= \frac{17-12}{8}$$

$= 0.63\text{ m/s}^2$

18 Explain the difference in the meaning of the terms:

a) speed

b) acceleration.

19 Explain what is meant by an acceleration of:

a) 2 m/s^2

b) -4 m/s^2

20 In the table shown, calculate the value of each missing quantity.

	Acceleration (m/s^2)	Change in velocity (m/s)	Time taken (s)
a)		40	5
b)		56	8
c)	9.8	39.2	
d)	2.4	96	
e)	8.2		25
f)	6.4		8.0

21 The velocity of a car increases by 20 miles per hour in five seconds. Calculate the acceleration of the car in miles per hour per second.

22 In the table shown, calculate the value of each missing quantity.

	Acceleration (m/s^2)	Initial velocity (m/s)	Final velocity (m/s)	Time for change (s)
a)		0	20	5
b)		32	64	8
c)		20	0	5
d)		45	20	4
e)	0.5	25		15
f)	9.8	0		8.4
g)	2.5		25	8.0
h)	−7		10	3
i)	−12	48	18	
j)	−10	35	13	

23 A car starts from rest and accelerates uniformly along a straight road. The acceleration of the car is $1.5\,m/s^2$. What is the speed of the car 8 s after starting?

24 An aircraft accelerates uniformly from rest to 36 m/s in 18 s. What is the acceleration of the aircraft?

25 A train travelling at 15 m/s accelerates uniformly to 35 m/s in 120 s. Calculate the acceleration of the train.

26 A small boat travelling at 10 m/s accelerates uniformly to rest in 25 s. Calculate the acceleration of the boat.

27 A lorry is accelerating uniformly at $3.0\,m/s^2$ along a straight road. How long did it take for the speed of the lorry to increase from 4.0 m/s to 16 m/s?

28 A car accelerates uniformly from rest to 15 m/s in 12 s. Calculate the acceleration of the car.

29 The speed of an aircraft as it lands on a runway is 60 m/s. The aircraft accelerates uniformly at $-8\,m/s^2$ and comes to rest. Calculate the time taken for the aircraft to come to rest after landing on the runway.

30 A racing car accelerates uniformly from 15 m/s to 27 m/s as it travels along a straight road. The acceleration of the car is $15\,m/s^2$. Calculate the time for this change in speed to occur.

Example question

The graph shows how the velocity of a car varies with time.

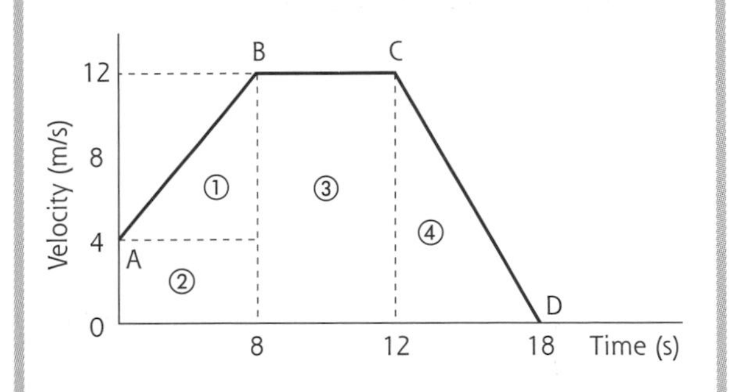

a) Describe the motion of the car for the three sections of the graph

b) Calculate:
i) the acceleration of the car during AB
ii) the acceleration of the car during CD.

c) Calculate the displacement of the car 18 seconds after starting.

Solution

a) AB constant acceleration; BC constant speed; CD constant negative acceleration (deceleration))

b) i) $u = 4\,\text{m/s}$
$v = 12\,\text{m/s}$
$t = 8\,\text{s}$

$$a = \frac{v - u}{t} = \frac{12 - 4}{8} = 1\,\text{m/s}^2$$

ii) $u = 12\,\text{m/s}$
$v = 0\,\text{m/s}$
$t = 6\,\text{s}$

$$a = \frac{v - u}{t} = \frac{0 - 12}{6} = -2\,\text{m/s}^2$$

c) Displacement = area under the graph
= area of triangle + area of rectangle + area of rectangle + area of triangle
= area of (1) + (2) + (3) + (4)
= $0.5 \times 8 \times (12 - 4) + 4 \times 8 + 12 \times 4 + 0.5 \times 12 \times 6$
= $32 + 32 + 48 + 36$
= 148 m

31 Draw a velocity–time graph to show the motion of a vehicle travelling at a constant velocity of 5 m/s for 6 s.

32 A bus accelerates uniformly from rest along a straight road. The speed of the bus after 5 s is 10 m/s. The bus then travels at this velocity for a further 4 s. The bus then accelerates uniformly for 6 s. The velocity of the bus is now 4 m/s. Draw a graph to show how the velocity of the bus varies with time.

33 The graphs show how the velocity of four different objects varies with time. Describe the motion of each object.

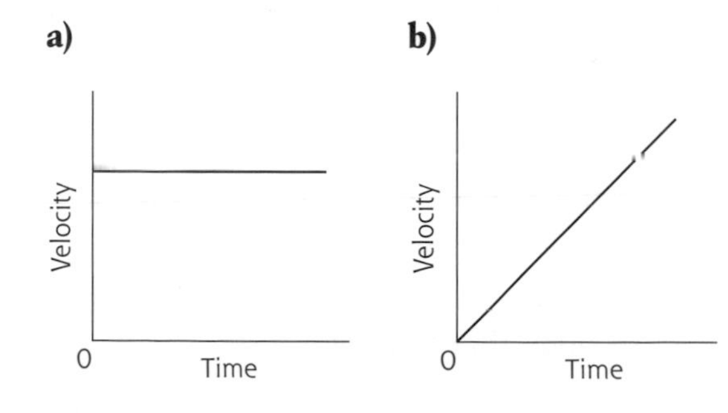

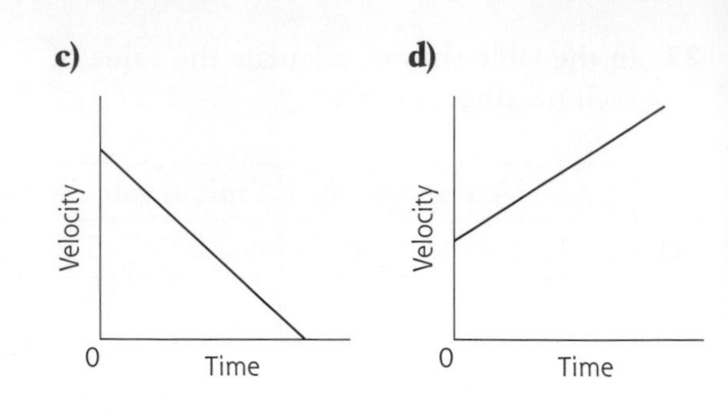

34 The graphs below show how the velocity of four different objects varies with time. Calculate the acceleration of each object.

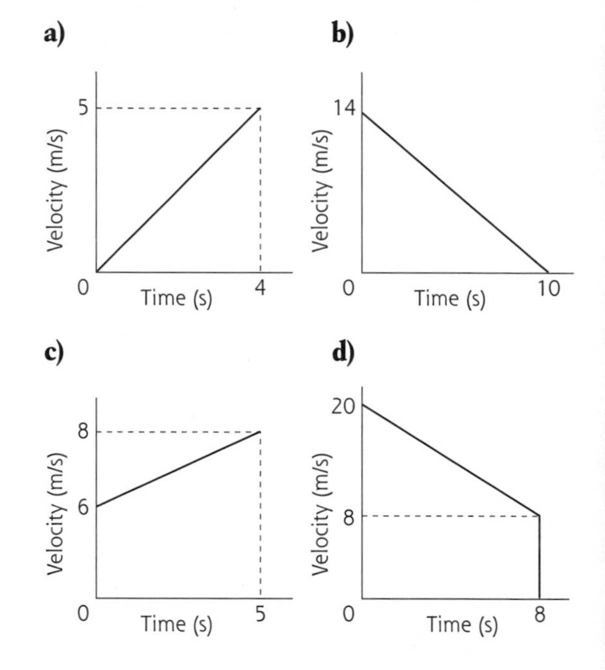

35 The graphs show how the velocity of four different objects varies with time. Calculate:

i) the displacement of each object

ii) the acceleration of each object.

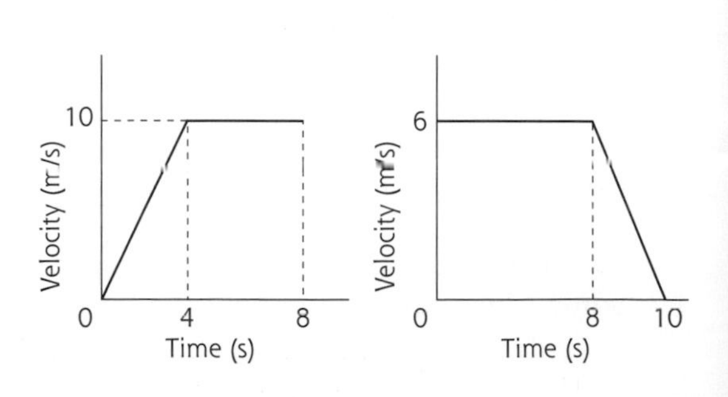

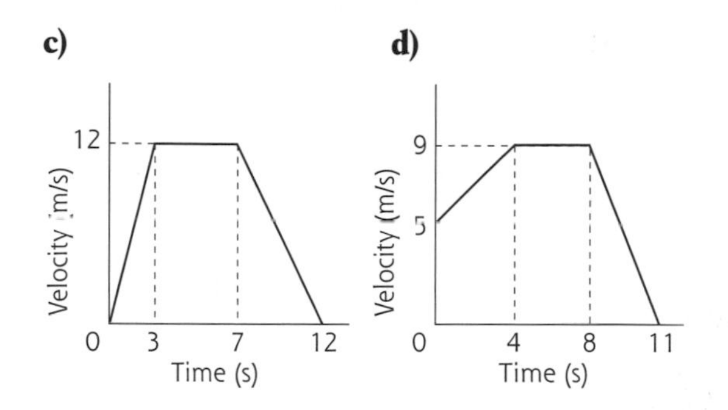

36 A ball is dropped onto a floor and rebounds without a change in speed back to the same height. The graph shows how the velocity of the ball varies with time. The ball hits the floor at A.

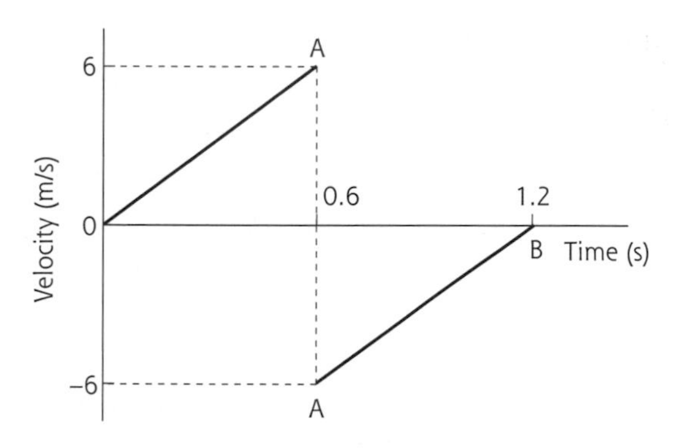

a) Describe the motion of the ball between:
i) OA
ii) AB.

b) Calculate the acceleration of the ball between:
i) OA
ii) AB.

c) Calculate the displacement of the ball 1.2 s after being dropped.

d) What was the initial height of the ball above the floor?

37 The graph shows how the velocity of a vehicle varies with time.

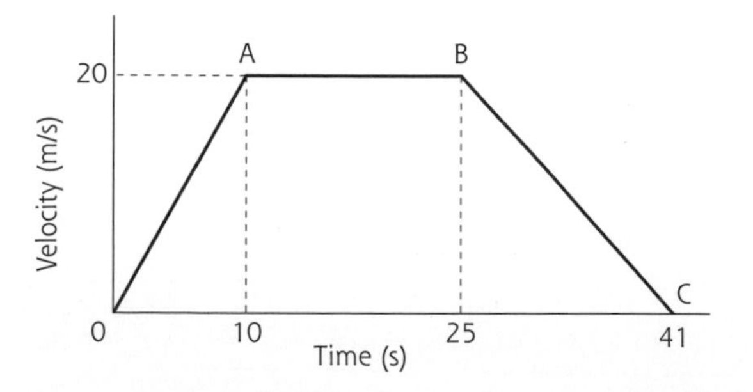

a) Describe the motion of the vehicle between:
i) OA
ii) AB
iii) BC.

b) Calculate the acceleration of the vehicle between:
i) OA
ii) AB
iii) BC.

c) Calculate the distance travelled by the vehicle.

d) Calculate the average speed of the vehicle.

38 A trolley accelerates down a slope. A student suggests two methods to measure the instantaneous speed of the trolley at point X.

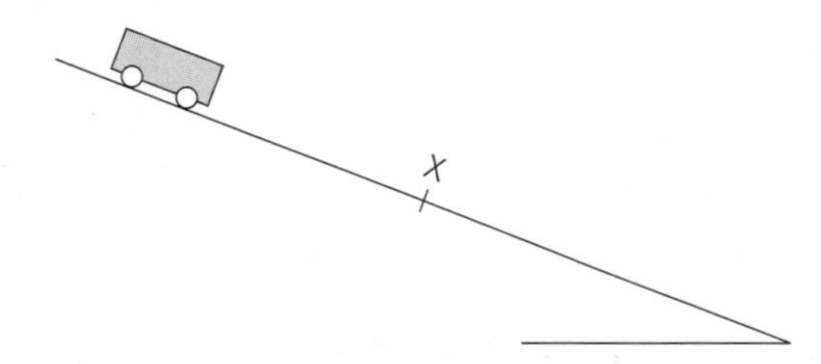

Method A – the student uses a stopwatch to time how long the length of the trolley took to pass point X on the slope.

Method B – the student uses a light gate connected to a timer to time how long the length of the trolley took to pass point X on the slope.

a) Explain why each method is liable to give a different instantaneous speed for the trolley at point X.

b) Which method is likely to give the best estimate of the instantaneous speed at point X?

Forces

- Forces can change the shape, speed and direction of travel of an object.
- A newton balance is used to measure the size of a force.
- Force is measured in newtons (N).
- Mass is the amount of matter in a substance but weight is a force and is the Earth's pull on the object.
- The weight per unit mass (i.e. weight/mass) is called the gravitational field strength.
- $W = mg$ where g is the gravitational field strength.
- Mass is measured in kilograms (kg), weight in newtons (N) and gravitational field strength in newtons per kilogram (N/kg).
- Acceleration due to gravity = gravitational field strength.
- The force of friction can oppose the motion of an object.
- It can be advantageous to increase the force of friction when you do not wish something to start moving or when you wish something that is moving to slow down e.g. car brakes.
- Force is a vector.
- Forces of the same size acting in opposite directions on an object are called balanced forces and are equivalent to no force at all.
- Newton's First law states that an object will remain at rest or travel at constant velocity unless acted on by an unbalanced force.
- Newton's Second Law is $F_{un} = ma$.
- The resultant force is the one force which can replace all the forces acting on an object and have the same effect.
- Projectile motion can be treated as two independent motions.
  - Horizontal motion – the object travels at constant velocity.
  - Vertical motion – the object has a constant downwards acceleration.

QUESTIONS

Example question

The gravitational field strength on the Earth is 10 N/kg. What is the weight of a person of mass 52 kg on Earth?

Solution

$W = mg$
$= 52 \times 10$
$= 520$ N

1 Describe three effects that a force can have on an object.

2 Describe how you would use a newton balance to measure the force applied to an object.

3 What does it mean that a mass of 5 kg has a weight of 50 N?

4 In the table shown, calculate the value of each missing quantity.

	Weight (N)	Mass (kg)	Gravitational field strength (N/kg)
a)		2.5	10
b)		17	10
c)	90		10
d)	220		10
e)		50	1.6
f)		72	4.0

5 What is the weight of a 7.0 kg mass on the Earth?

6 On the Earth, an object is suspended vertically from a newton balance. The balance is held stationary. The reading on the balance is 6.0 N. What is the mass of the object?

7 An astronaut has a mass of 60 kg on Earth. She lands on the Moon where the gravitational field strength is 1.6 N/kg.

a) What is her weight on the Moon?

b) What is meant by the gravitational field strength is 1.6 N/kg?

c) What is the astronaut's weight on the Earth?

d) **i)** What is the astronaut's mass on the moon?
ii) Explain your answer.

8 A space probe of mass 300 kg lands on Mars. The gravitational field strength on Mars is 3.8 N/kg. Calculate the weight of the probe on the surface of Mars.

9 What is the difference between mass and weight?

10 A car is travelling due north. In which direction does:

a) the engine force act?

b) the force of friction act?

11 Describe one everyday situation where it is useful to:

a) increase the force of friction

b) decrease the force of friction.

12 A car is travelling along a road at a constant speed of 15 m/s.

a) Name the two forces acting on the car.

b) What can you state about the size of these forces?

13 An aircraft is travelling at a height of 10 000 m at a constant speed of 300 m/s. The engine produces a forward thrust force of 280 kN.

a) Name the two forces acting on the aircraft horizontally.

b) What is the size of both these forces?

14 A car is travelling at constant speed. The engine force and force of friction are the only forces acting on the car. Draw a labelled diagram showing the forces acting on the car. Indicate on your diagram the direction of travel of the car.

15 A lampshade of mass 1.2 kg is suspended from the ceiling by a flex.

a) Calculate the weight of the lampshade.

b) **i)** What is the size of the tension (the force) in the flex?
ii) In which direction does the tension act? Explain your answer.

16 Explain the motion of each of the following objects in terms of the forces acting on them:

a) a parachutist moving towards the ground at constant speed

b) a clock which is at rest on a table

c) a cyclist travelling along a level road at constant velocity.

17 Explain, in terms of the horizontal force or forces acting on a person, why seat belts are used in cars and buses.

Example question

A car of mass 1250 kg accelerates uniformly along a road at 4.5 m/s². The engine force is 7000 N.

Calculate the frictional force acting on the car.

Solution

Accelerating force $F_{un} = ma$
$= 1250 \times 4.5 = 5625$ N

Total force acting is 7000 N

F_{un} = total force − frictional force
5625 = 7000 − frictional force

frictional force = 7000 − 5625
= 1375 N

18 For the objects shown below calculate the unbalanced force acting on each one.

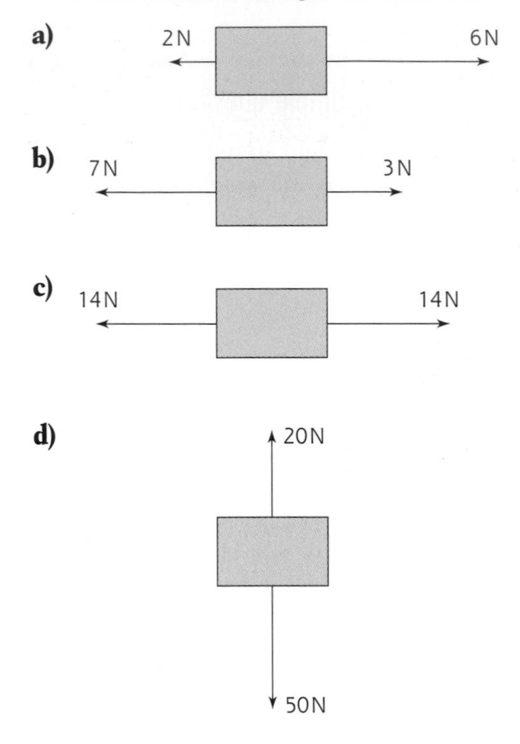

19 In the table shown, calculate the value of each missing quantity.

	Unbalanced force (N)	Mass (kg)	Acceleration (m/s²)
a)		15	6
b)		0.8	2.5
c)	150	37.5	
d)	240	8	
e)	40		5
f)	96		3.6

20 The mass of a car is 1500 kg. Calculate the unbalanced force required to accelerate the car at 2 m/s².

21 A braking force of 6.6×10^4 N is applied to a lorry of mass 4.0×10^4 kg. Calculate the acceleration of the lorry.

22 An unbalanced force of 96 kN acts on a train. The acceleration of the train is 1.2 m/s².

Calculate the mass of the train.

23 In the table shown, calculate the value of each missing quantity.

	Force applied (N)	Frictional force (N)	Mass (kg)	Acceleration (m/s^2)
a)	25		4	2.5
b)	40		3.5	8
c)	50	10	5	
d)	78	18	20	
e)	120	24		4
f)	72	22		10

24 The mass of a trolley is 0.5 kg. A force of 2.0 N is applied to the trolley. The acceleration of the trolley is $3.0\,m/s^2$.

a) Calculate the unbalanced force acting on the trolley.

b) Calculate the size of the frictional force acting on the trolley.

25 During take-off, the engine of an aircraft provides a constant thrust. The mass of the aircraft is 1200 kg. The acceleration during take-off is $2.7\,m/s^2$. A constant frictional force of 500 N acts on the aircraft during the take-off.

a) Calculate the unbalanced force on the aircraft during take-off.

b) Calculate the thrust of the engine of the aircraft.

26 During the flight of a rocket:

a) the mass of the rocket decreases as fuel is used up

b) there is a reduction in the engine thrust as booster engines stop firing.

Describe the effect that each of these changes has on the acceleration of the rocket.

Example question

A ball rolls off the end of a table at a constant speed of 3.2 m/s. The ball takes 0.23 s to reach the ground. The effect of air resistance can be ignored.

a) Calculate the horizontal distance travelled by the ball.

b) Calculate the vertical velocity of the ball just before it reaches the ground.

Solution

a) $d = vt = 3.2 \times 0.23 = 0.74\,m$

b) $v = u + at = 0 + 10 \times 0.23 = 2.3\,m/s$

In the following questions the effect of air resistance can be ignored.

27 A ball rolls along a bench top and falls off the end. The ball follows a curved path as it falls to the ground.

Describe:

a) the horizontal motion of the ball

b) the vertical motion of the ball.

28 A ball is thrown horizontally off a building at a speed of 18 m/s. The ball takes 4.5 s to reach the ground.

a) Calculate the horizontal distance travelled by the ball.

b) Calculate the vertical speed of the ball just before it hits the ground.

c) Show that the average vertical speed of the ball during that journey is 22.5 m/s.

d) Calculate the vertical height of the building.

29 At a fireworks display a rocket is fired horizontally to try and strike a target. The graphs show how the horizontal and vertical speeds of the rocket vary with time.

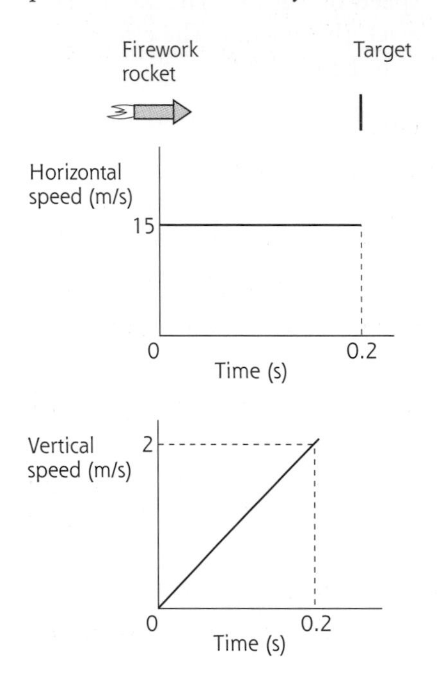

a) Calculate the horizontal distance travelled by the rocket.

b) The vertical height that the rocket has to drop to hit the target is 0.75 m. Show whether or not the rocket struck the target.

30 A trapeze artist at the circus swings out horizontally at a constant speed and then drops to the next swing below him. The time taken to fall to the next swing is 0.6 s.

a) What is his vertical velocity on hitting the next swing?

b) What is his average vertical speed?

c) What vertical distance does he travel?

d) The horizontal distance between the swings is 7.2 m. Calculate the horizontal speed of the trapeze artist.

31 A bowling ball of mass 5 kg rolls off the end of a bench. The graphs show how the horizontal and vertical speeds of the ball vary with time.

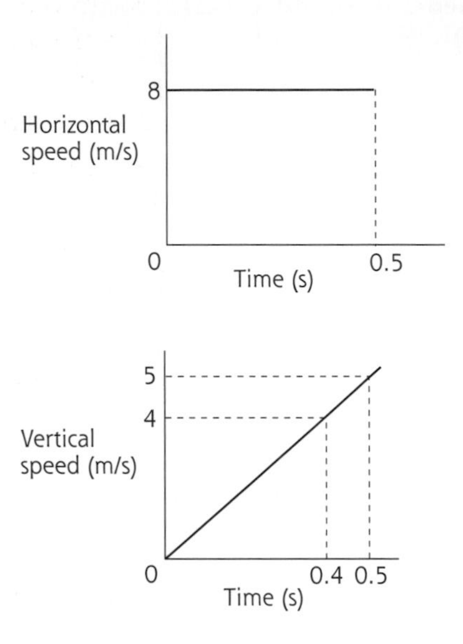

a) Calculate the horizontal distance travelled by the ball.

b) Calculate the vertical distance travelled by the ball.

c) Calculate the resultant velocity of the ball 0.4 s after it rolled off the bench.

d) A different ball of mass 7 kg rolls off the bench with the same horizontal speed.

What differences, if any, will be seen in both graphs? Explain your answer.

2.3 Momentum and energy

- Newton's third law states that forces always act in pairs which are equal in size but opposite in direction and the forces act on different objects.
- Momentum = mass × velocity.
- Momentum is a vector quantity and the unit is kg m/s.
- Total momentum before a collision = total momentum after, in the absence of external forces, for example friction.
- Work done = force applied × distance moved by the force.
- Force is measured in newtons (N), distance in metres (m) and work done in joules (J).
- The work done in lifting an object through a vertical height is equal to the gain in gravitational potential energy of the object.
- Mass is measured in kilograms (kg), gravitational field strength in newtons per kilogram (N/kg), vertical height in metres (m) and gravitational potential energy in joules (J).
- Change in gravitational potential energy, $E_p = mgh$.
- The work done in increasing the speed of an object is equal to the gain in kinetic energy.
- Mass is measured in kilograms (kg), speed in metres per second (m/s) and kinetic energy in joules (J).
- Kinetic energy, $E_k = \frac{1}{2} mv^2$.
- Energy can be changed from one form to another i.e. the amount of energy does not change – this is known as the Conservation of Energy.
- Power $P = \frac{\text{work done}}{\text{time}} = \frac{\text{energy transferred}}{\text{time}}$.
- Work done is measured in joules (J), time in seconds (s) and power in watts (W).
- Efficiency $= \frac{\text{useful work output}}{\text{total work input}} \times 100\%$.

QUESTIONS

Example question

A car of mass 1200 kg is travelling along a straight road at 15 m/s. The car collides with a stationary car of mass 1000 kg. After the collision the two cars lock together. Calculate the speed of the two cars after the collision.

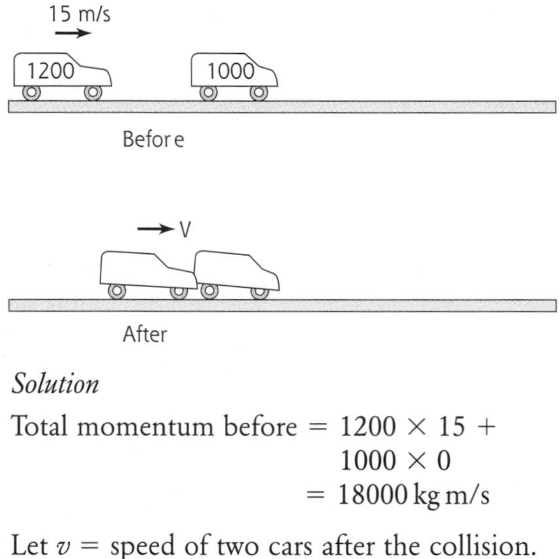

Solution

Total momentum before = 1200 × 15 + 1000 × 0
= 18000 kg m/s

Let v = speed of two cars after the collision.
Total momentum after = (1200 + 1000)v
= 2200v

From conservation of momentum:
Total momentum before = Total momentum after

$$2200v = 18000$$
$$v = 8.2\text{ m/s}$$

The speed of the vehicles after the collision is 8.2 m/s.

1. A car of mass 1000 kg is travelling at 12 m/s along a straight road. It collides and sticks to a stationary car of mass 1500 kg. Calculate the combined speed of the two cars after the collision.
2. A fully laden shopping trolley of mass 30 kg moves in a straight line at 6 m/s. The trolley collides into an empty trolley of mass 7 kg, which is stationary. Calculate the combined speed of the two trolleys after the collision.
3. A pupil of mass 50 kg is running at 8 m/s when he jumps on a stationary skateboard. After the collision the pupil and skateboard move off at 7.4 m/s. Calculate the mass of the skateboard.
4. A lorry of mass 1800 kg is travelling at 12.5 m/s along a straight road when it collides with another vehicle of mass 1200 kg, which is at rest. The vehicles join together. Calculate their combined speed after the collision.
5. A space shuttle is moving at 15 m/s. The shuttle has a mass 10 000 kg. It docks with a stationary space station of mass 50 000 kg. Calculate the speed of the shuttle and station after docking.
6. A car has a mass of 1200 kg and is travelling along a straight road. The speed of the car is 16 m/s. It collides with a car of mass of 1050 kg which is at rest. The car of mass 1200 kg continues in a straight line at 7 m/s. Calculate the speed of the car of mass 1050 kg.
7. A lorry of mass 1400 kg is moving at 12.5 m/s in a straight line. It collides with a stationary car of mass 1000 kg. After the collision the lorry goes on in a straight line at 5.5 m/s. Calculate the speed of the car.
8. A boat of mass 800 kg is travelling through the water at 11.5 m/s. The boat collides with a stationary log of mass 60 kg. The boat and log stick together. Calculate the combined speed of the boat and the log after the collision.
9. A spacecraft of mass 2500 kg is moving at 17 m/s when it docks with a stationary space probe of mass 1250 kg. Calculate the combined speed of the spacecraft and probe after docking.
10. A racing car of mass 1300 kg is travelling in a straight line at 22 m/s. The car collides with a stationary barrier. The mass of the barrier is 125 kg. Calculate the combined speed after the collision if the car and barrier stick together.

Example question

A trolley is pushed along a straight path with a force of 45 N. The trolley is pushed for a distance of 22 m. Calculate the work done.

Solution

Work done = force × distance
= 45 × 22
= 990 J

11 A car, which uses petrol as a fuel, is travelling along a straight, horizontal road. Describe the main energy change which takes place when the car:

a) accelerates

b) moves at constant speed

c) brakes.

12 A petrol engine car is travelling along a straight road. Describe the main energy change which takes place when the car travels at constant speed:

a) up a slope

b) down a slope with the engine switched off.

13 A ball is released from a height of 1 m and falls to the ground. After hitting the ground it rebounds to a height of 0.75 m. The effects of air resistance can be neglected.

State the energy change which takes place

a) when the ball falls to the ground

b) when the ball is in contact with the ground

c) when the ball rebounds to a height of 0.75 m.

14 In the table shown, calculate the value of each missing quantity.

	Work done (J)	Force (N)	Distance (m)
a)		25	7
b)		30	4.5
c)	250	12.5	
d)	700	35	
e)	12 000		200
f)	60 000		1 500

15 A box is pulled 3 m along the floor by a horizontal force of 50 N. Calculate the work done by the force.

16 A boy is pedalling his bicycle along a horizontal level road. In covering a distance of 40 m, the bicycle gains 3000 J of energy. Calculate the minimum force exerted by the boy on the pedals.

17 When braking, the force exerted by the brakes of a car is 1500 N. During this time the kinetic energy of the car decreases by 45 000 J. Calculate the distance travelled by the car during braking.

18 The engine of a train exerts a force of 8 kN and does 2×10^6 J of work. Calculate the minimum distance travelled by the train.

19 A filing cabinet of mass 100 kg is pushed 0.25 m so that it is against a wall. During this movement 60 J of work are done. Calculate the horizontal force the person exerted on the cabinet.

20 A car of mass 1500 kg travels 0.5 km along a horizontal road at a constant speed. The engine exerts a constant force of 3000 N.

a) State the size of the frictional force which acts on the car.

b) Calculate the work done against friction.

Example question

A kettle uses 5.04×10^5 J when heating water. The kettle is switched on for 3.5 minutes.

Calculate the power of the kettle.

Solution

Energy = 5.04×10^5 J

Time = 3.5 minutes = 210 s

$$\text{Power} = \frac{\text{Energy transferred}}{\text{time}}$$

$$= \frac{5.04 \times 10^5}{210} = 2400\ \text{W}$$

21 In the table shown, calculate the value of each missing quantity.

	Power (W)	Work done (J)	Time (s)
a)		240 000	100
b)		1 500	300
c)	1 100		480
d)	1 250		750
e)	150	3 750	
f)	7 000	420 000	

22 The motor of a blender does 4.2 kJ of work in 30 s. Calculate the power of the motor.

23 The power of an electric motor is 12W. How much work is done by the motor in two minutes?

24 The power of a machine is 0.96 kW. Calculate the time taken to transfer 4800 J of energy.

Example question

A car has a mass of 800 kg. It is travelling along a straight road at a constant speed of 5 m/s.

Calculate the kinetic energy of the car.

Solution

$E_k = \frac{1}{2} mv^2$

$= \frac{1}{2} \times 800 \times (5)^2$

$= 10\,000\,\text{J}$

25 In the table shown, calculate the value of each missing quantity.

	Kinetic energy (J)	Mass (kg)	Speed (m/s)
a)		3	2
b)		5	8
c)	62.5	5	
d)	490	20	
e)	1 000		5
f)	80		8

26 A girl on skates is moving at a speed of 1.5 m/s. The mass of the girl is 40 kg. How much kinetic energy does the girl have?

27 A football travelling at 5 m/s has 27 J of kinetic energy. What is the mass of the football?

28 A vehicle has 0.4 J of kinetic energy. The vehicle has a mass of 0.2 kg. Calculate the speed of the vehicle.

29 A ball bearing has a mass of 60 g and is propelled across a floor at a speed of 200 mm/s. Calculate the kinetic energy of the ball bearing.

30 When travelling at a speed of 0.5 m/s a toy car has 25 mJ of kinetic energy. Calculate the mass of the car.

31 A car has a mass of 1 500 kg and has 675 kJ of kinetic energy. Calculate the speed of the car.

32 In the table shown, calculate the value of each missing quantity.

	Change in gravitational potential energy of object	Mass of object	Vertical height moved by object
a)		5 kg	3.5 m
b)		500 g	700 mm
c)	100 J		2.5 m
d)	20 J		0.75 m
e)	300 J	5 kg	
f)	80 J	500 g	

33 How much gravitational potential energy does a package of mass 6.0 kg gain when lifted 1.5 m?

34 A bag of mass 6 kg is lifted onto a shelf. The bag gains 72 J of gravitational potential energy. Calculate the vertical height the bag is lifted through.

35 A brick drops 3.6 m onto the ground. The brick loses 90 J of gravitational potential energy. Find the mass of the brick.

36 A sack of mass 24 kg is pulled 5 m along a level floor by a horizontal force of 100 N. The sack is then lifted 0.8 m onto a workbench. Calculate the total work done on the sack.

37 A ball of mass 1.5 kg is dropped from a window. It drops 5 m onto the ground below. The effects of air resistance may be ignored.

a) Calculate the loss in gravitational potential energy of the ball.

b) What is the kinetic energy of the ball just before it hits the ground?

c) Find the speed of the ball just before it hits the ground.

38 A swimmer of mass 45 kg steps off a diving board and falls, from rest, into the water. The swimmer is travelling at 8 m/s just before she hits the water. The effect of air resistance may be ignored.

a) Calculate the kinetic energy of the swimmer just before hitting the water.

b) How much gravitational potential energy does the swimmer lose as she falls into the water?

c) Calculate the height of the diving board above the water.

39 A crane lifts a bucket through a vertical height of 20 m to the top of a building in 30 s. The mass of the bucket is 200 kg. Calculate:

a) the gravitational potential energy gained by the bucket during this time

b) the minimum power output of the crane during this time.

40 A lift travels 35 m vertically up a lift shaft at constant speed for a time of 50 s. The lift has a total mass of 5 000 kg.

a) Calculate the gravitational potential energy gained by the lift.

b) What is the minimum power output of the lift motor?

Example question

The input power of a machine is 1.5 kW. The output power is 1.0 kW.

Calculate the efficiency of the machine.

Solution

$$\text{Efficiency} = \frac{\text{power output}}{\text{power input}} \times 100\%$$

$$= \frac{1000}{1500} \times 100\%$$

$$= 67\%$$

41 In the table shown, calculate the value of each missing quantity.

	Efficiency %	Energy input	Energy output
a)		500 J	400 J
b)		140 kJ	70 kJ
c)	30	750 J	
d)	28	560 J	
e)	75		3.6 MJ
f)	40		160 kJ

42 In the table shown, calculate the value of each missing quantity.

	Efficiency %	Power input	Power output
a)		500 W	400 W
b)		140 kW	70 kW
c)	30	750 W	
d)	28	560 W	
e)	75		3.6 MW
f)	40		160 kW

43 The power input to a power station is 1680 MW. The electrical power output of the power station is 800 MW.

Calculate the efficiency of the power station.

44 Water in a hydroelectric power station falls through a vertical height of 50 m. The water passes through the turbine at the rate of 1500 kg every second. The power station is able to convert 80% of the water's energy into electrical energy.

a) Calculate the loss in gravitational potential energy of the water every second.

b) How much electrical energy does the power station produce every second?

45 A small wind turbine is used to charge batteries. The wind turbine is able to convert 30% of the wind's energy into electrical energy. During part of the day the wind turbine produces 6 W of electrical power.

a) Calculate the power input to the wind turbine during this time.

b) How much electrical energy is produced in one hour?

46 A house has solar cells installed on its roof to provide electrical power. The solar cells are installed in panels which have a total area of 5 m^2. On a dull day the panels receive 800 W on each square metre.

a) Calculate the total power received by the panels on a dull day.

b) The panels produce 320 W of electrical power. Calculate the efficiency of the system.

2.4 Heat

- Equal masses of different substances require different amounts of energy to change their temperature by 1 °C.
- The energy absorbed or released by a substance, E_h, is measured in joules (J), specific heat capacity, c, is measured in joules per kilogram per degree celsius (J/kg °C), mass, m, is measured in kilograms (kg) and the change in temperature, ΔT, is measured in degrees celsius (°C).
- Energy absorbed or released = specific heat capacity × mass × change in temperature, i.e. $E_h = cm\Delta T$

  This equation is used whenever there is a change in temperature of the substance.
- A specific heat capacity, c, of 100 J/kg °C means that 100 J of energy is required to change the temperature of 1 kg of the substance by 1 °C
- A change of state occurs when a solid changes into a liquid (or a liquid changes to a solid) or a liquid changes to a gas (or a gas changes to a liquid).
- There is no change in temperature when a change of state occurs.
- During a change in state, the energy absorbed or released by the substance, E_h, is measured in joules (J), mass, m, is measured in kilograms (kg) and the specific latent heat of fusion or vaporisation, l, is measured in joules per kilogram (J/kg).
- Energy absorbed or released = mass × specific latent heat, i.e. $E_h = ml$

  This equation is used whenever there is a change in state of the substance.

QUESTIONS

You may need to refer to the data sheet on p. 98 for some of the information required in the questions which follow.

1 In the following sentences the words represented by the letters A, B, C and D are missing.

A thermometer measures the ___A___ of an object. It uses the ___B___ scale in units called degrees celsius (°C). When energy moves from one object to another, ___C___ from the warm object moves to the cooler object by the processes of ___D___, convection and radiation.

Match each letter with the correct word from the box.

celsius **conduction** **fahrenheit** **heat** **temperature**

2 Copper has a specific heat capacity of 386 joules per kilogram per degree celsius. Aluminium has a specific heat capacity of 902 joules per kilogram per degree celsius.

You are given blocks of equal masses of both metals and asked to raise the temperature of each one by 1 °C. Which block, copper or aluminium, would require most energy to do this? Give a reason for your answer.

Example question

Water at an initial temperature of 18 °C is heated until it reaches 45 °C. The mass of the water is 1.5 kg. Calculate the energy absorbed by the water.

Solution

$E_h = cm\Delta T = 4180 \times 1.5 \times (45 - 18)$
$= 169\,290 = 1.69 \times 10^5$ J

3 Calculate the amount of energy required to increase the temperature of:

a) 3 kg of water by 5 °C

b) 8 kg of water by 3 °C

c) 1.5 kg of water by 80 °C.

4 In the table shown, calculate the value of each missing quantity.

	E_h (J)	c (J/kg °C)	m (kg)	ΔT (°C)
a)		150	2	5
b)	2 000		5	2
c)	30 000	500		120
d)	20 900	4 180	0.1	

5 Calculate the amount of energy required to increase the temperature of 0.5 kg of aluminium by 20 °C.

6 A kettle is filled with 0.4 kg of water at a temperature of 15 °C. Calculate the minimum amount of energy required to raise the temperature of the water to its boiling point.

7 An electric heater has a power rating of 50 W. The heater is used to heat a copper block of mass 1 kg. The heater is switched on for 193 s. The temperature of the block rises from 15 °C to 35 °C in this time.

a) Show that the heater transfers 9650 J of energy in 193 s.

b) Calculate the energy absorbed by the copper block during this time.

c) Calculate the efficiency of heating the block in this way.

8 A small immersion heater is used to boil water in a beaker. The beaker holds 0.5 kg of water. The initial temperature of the water is 20 °C. The immersion heater has a power rating of 500 W. Calculate the minimum time taken to raise the temperature of the water to its boiling point.

9 The power rating of an immersion heater is 1.0 kW. The heater is used to heat a liquid. The mass of the liquid is 2.5 kg. The initial temperature of the liquid is 20 °C. The heater is switched on for two minutes and the temperature of the liquid rises to 44 °C. Calculate the specific heat capacity of the liquid.

10 In the following sentences the words or formulae represented by the letters A, B, C, D, E and F are missing.

The equation E_h = ___A___ is used whenever a solid, liquid or gas of mass, *m*, undergoes a change in temperature, ΔT. The symbol *c* represents the ___B___ of the substance.

The equation E_h = ___C___ is used whenever a change in state occurs where *m* is the mass of the substance which changes state. When a solid changes to a liquid at its melting point, *l* represents the specific latent heat of ___D___. When a liquid changes to a gas at its boiling point, *l* represents the specific latent heat of ___E___. In both cases there is no change in ___F___ and energy is gained (or released) by the substance during the change of state.

Match each letter with the correct word or formula from the box.

$cm\Delta T$ ml IV   fusion
specific heat capacity   temperature
vaporisation

Example question

Ice cubes are put in a glass of water. The ice cubes reach 0 °C. They have a mass of 25 g. Calculate the energy absorbed by the ice cubes as they melt into water at constant temperature.

Solution

$E_h = ml = 3.34 \times 10^5 \times 25 \times 10^{-3} = 8350\,J$

11 In the table shown, calculate the value of each missing quantity.

	E_h (J)	m (kg)	l (J/kg)
a)		2.5	200 000
b)	1.67×10^6		334 000
c)	50 000	0.4	
d)		0.03	2.26×10^6
e)	105 000		150 000

12 How much energy is required to melt the following masses of ice at 0 °C:

a) 1 kg

b) 2 kg

c) 5 kg

d) 0.4 kg

13 How much energy is required to vaporise the following masses of water at 100 °C:

a) 1 kg

b) 2 kg

c) 5 kg

d) 0.4 kg

14 The power rating of a heater is 1000 W. The heater raises the temperature of a sample of water to its boiling point. The mass of the sample is 1 kg. The heater is left switched on for a further 113 s. Calculate the maximum mass of water that changes state in this time.

15 Some ice cubes are placed in a glass of water. After a few minutes, the temperature of the mixture reaches 0 °C.

The mixture now absorbs heat from the surroundings at a steady rate of 10 J/s.

Calculate the mass of ice that changes state in six minutes.

16 A scald from steam, at 100 °C, is often much worse than a scald from boiling water.

a) Calculate the energy released by 0.1 kg of boiling water as it cools to 30 °C, the temperature of the skin.

b) i) Calculate the energy released by 0.1 kg of steam at 100 °C as it changes to water at 100 °C.
ii) The water produced in **b) i)** must still cool down to 30 °C. Calculate the total amount of energy released by the steam as it cools.

c) Explain clearly why a scald from steam at 100 °C is worse than a scald from the same mass of boiling water.

17 The element of a small heating coil is placed in a container of liquid nitrogen. The nitrogen is at its boiling point. The element is switched on and the nitrogen absorbs 1000 J of energy. This changes 5 g of the liquid nitrogen into nitrogen gas at constant temperature.

Calculate the specific latent heat of vaporisation of nitrogen.

18 Water is poured into an ice tray and then placed in a freezer. The total mass of water in the tray is 0.30 kg.

a) It takes five minutes to reduce the temperature of the water from 20 °C to 0 °C.

Show that energy is removed from the water at a rate of 83.6 J/s.

b) Assuming that energy continues to be removed at 83.6 J/s, how long it will take for the water at 0 °C to change into ice at 0°C?

19 Give **two** examples of applications which involve a change of state.

Exam-style questions

1 a) Jenny cycles to and from school. Her journey takes her along a small stretch of straight road, past her friend Donna's house. Donna decides to measure Jenny's average speed along the straight road. She uses a stopwatch and a measuring tape.

Describe how Donna could measure Jenny's average speed as she travels along the straight road.

b) During a part of her journey, Jenny has to carry her bicycle up a flight of stairs. There are 10 stairs, each stair being 80 mm high. Her bicycle has a mass of 15 kg.
i) What is the minimum force required to lift the bicycle?
ii) Calculate the minimum work done in raising the bicycle up the flight of 10 steps.

2 A ship of mass 250 000 kg is towed at a constant speed of 1.5 m/s by a tug. The tug exerts a constant force of 1200 N on the towing cable when pulling the ship.

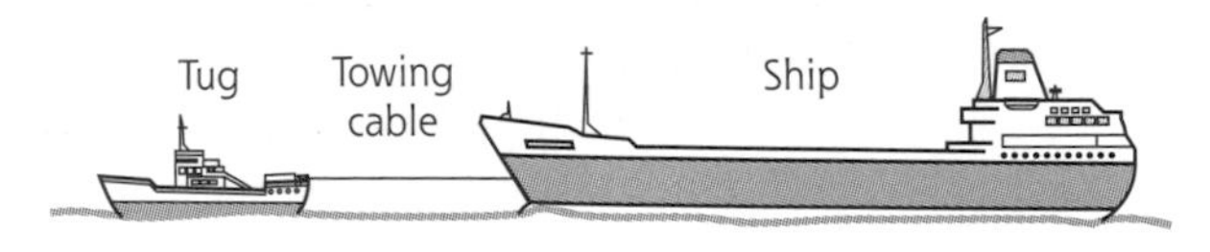

a) What is the size of the resistive force of the water opposing the motion of the ship? Explain your answer.

b) The towing cable is then released and the ship comes to rest. The resistive force of the water on the ship remains constant as the ship comes to rest.
i) Calculate the acceleration of the ship.
ii) Calculate the time for the ship to come to rest after the towing cable was released.
iii) Draw a graph to show how the speed of the ship varies with time from the instant the towing cable is released until the ship comes to rest.
iv) How far does the ship travel after the towing cable is released?

3 Bales of straw are raised at constant speed by an elevator from a platform, 0.5 m above the ground, onto a stack. The stack is 5.5 m above the ground. The bales have a mass of 25 kg.

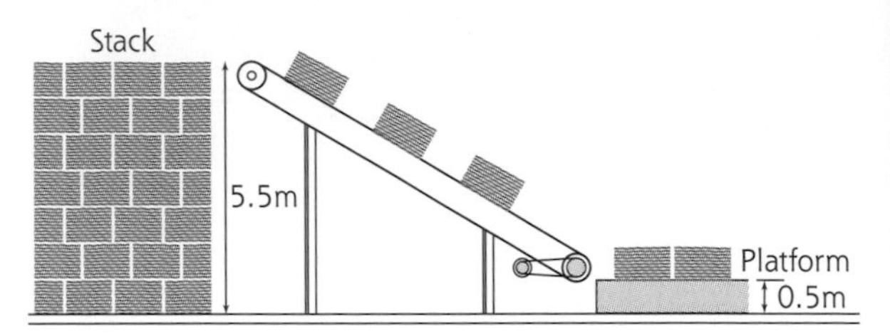

a) How much gravitational potential energy does a bale gain in being lifted from the platform to the stack?

b) The elevator can raise 20 bales of straw to the stack in five minutes.
i) Calculate the minimum power output of the electric motor of the elevator during this time.
ii) Give one reason why the input power to the electric motor is greater than the value calculated in **b) i)**.

4 The driver of a train sees that the signal ahead is at red and applies the train's brakes. The graph shows how the speed of the train varies with time from the instant the signal is seen by the driver.

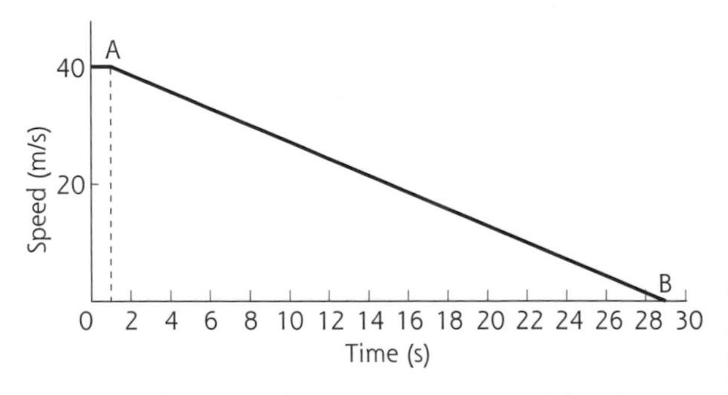

a) What was the reaction time of the driver?

b) Calculate the acceleration of the train between A and B.

c) The train is 620 m away from the signal when the driver sees it. Does the train stop in this distance? You must clearly show the working which leads to your answer.

5 The diagram below shows a pendulum bob at point X, its rest position. The pendulum bob has a mass of 0.2 kg.

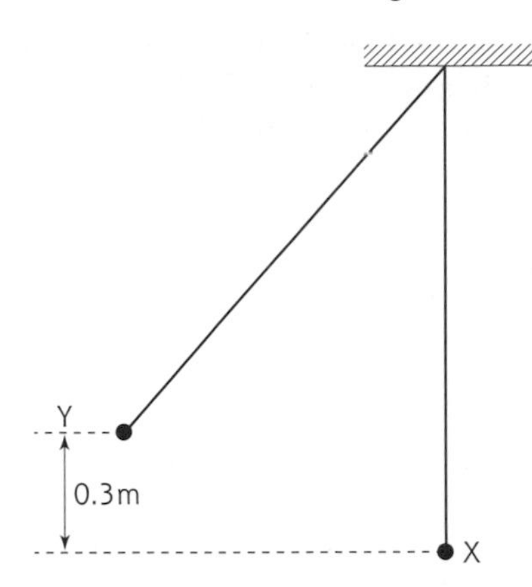

The pendulum bob was then pulled to point Y. Point Y is 0.3 m above the rest position.

a) Find the gain in gravitational potential energy of the bob when it is moved from point X to point Y.

b) The pendulum bob is released from point Y and swings to and fro until it comes to rest.
i) Describe the energy changes which takes place as the pendulum bob swings from point Y to point X.
ii) Show that the maximum possible speed of the pendulum bob is 2.45 m/s. State any assumption you make in your calculation.

6 A girl starts from rest at point P on a ski slope. She accelerates uniformly down the slope to point Q. She then slows down as she moves along a level slope and stops at point R. Point P is 20 m vertically above point Q. The mass of the skier is 45 kg.

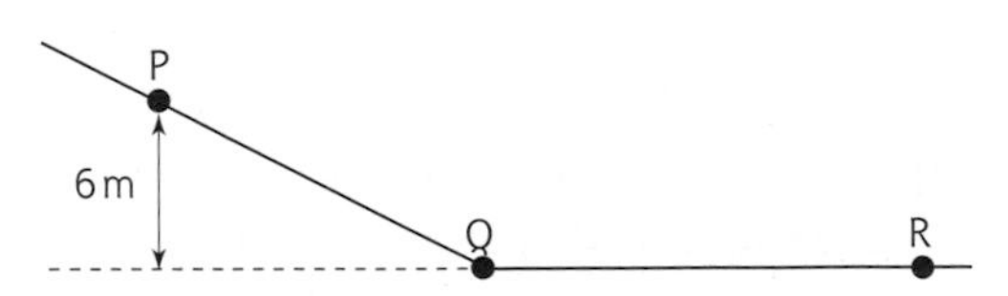

a) Calculate the change in gravitational potential energy of the girl as she moves from point P to point Q.

b) Assuming that all her gravitational potential energy is transferred to kinetic energy, calculate her speed at point Q.

c) As the girl passes point Q, there is a constant frictional force of 15 N acting on her. This force causes her to slow down to rest during the level section between points Q and R.

Calculate the distance travelled by the girl as she skis along the level section between points Q and R.

7 During a sailing competition, a boy competes in a race. The graph shows how the speed of the boat varies with time during the race.

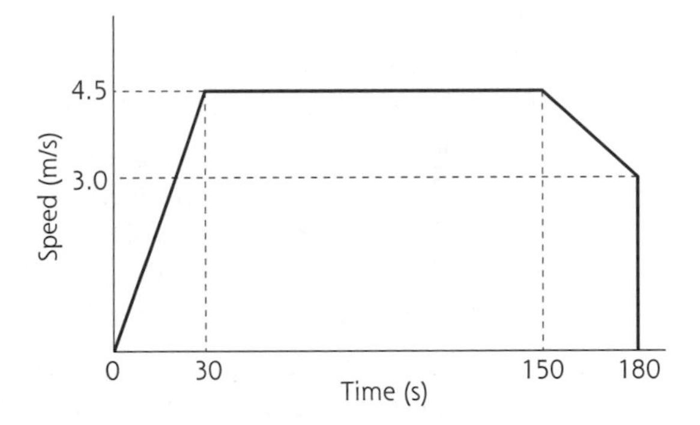

a) **i)** Find the acceleration of the boat during the first 30 seconds of the race.
ii) Calculate the distance travelled by the boat during the race.
iii) What is the average speed of the boat during the race?

b) The diagram below shows the horizontal forces acting on the boat during the race.

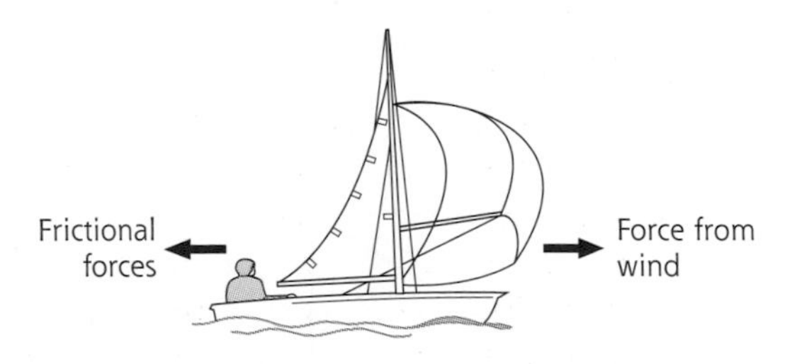

i) During the first 30 seconds of the race are these forces balanced or unbalanced? Justify your answer.
ii) Between the times of 30 s and 150 s during the race are these forces balanced or unbalanced? Justify your answer.

8 a) A ship of mass 14 000 kg is travelling south at a velocity of 12 m/s. A wind blows from east to west with a velocity 6 m/s. Calculate the resultant velocity of the ship.

b) The wind stops blowing and the ship continues at a constant speed of 10 m/s in a straight line. The frictional force exerted by the water is 120 N. What is the forward force acting on the ship? Explain your answer.

c) The ship is still travelling at 10 m/s in a straight line when it collides with a smaller boat. This boat has a mass of 3000 kg and is at rest. What is the combined speed of the two vessels if they stay together after collision?

9 A ball, X, of mass 1.5 kg is rolling along a flat horizontal rooftop at a constant speed of 7.5 m/s. It collides with another ball, Y, of mass 1 kg, which is at rest. After the collision ball X moves off in the original direction at 4 m/s.

a) Calculate the speed of ball Y after the collision.

b) Ball Y then falls off the roof top and hits the ground 4.2 s later.
i) Calculate the horizontal distance travelled by the ball.
ii) Calculate the vertical speed of the ball just before it hits the ground.
iii) Calculate the height of the building.

10 A ball is dropped vertically from a window of a house onto the ground below. The graph shows how the velocity of the ball varies with time for the first 0.8 s of its motion.

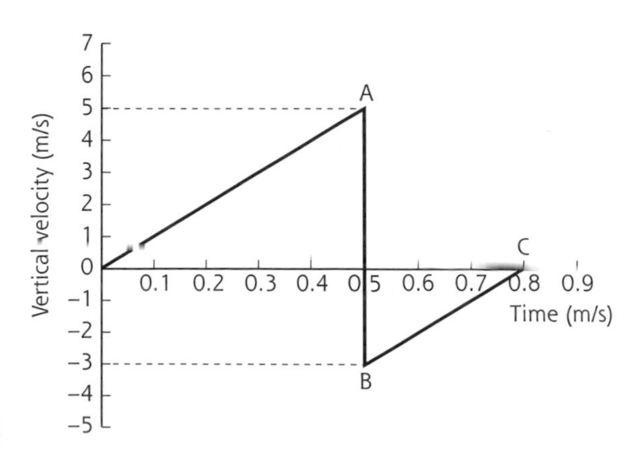

a) Describe the motion of the ball during
i) OA
ii) BC.

b) Calculate the height of the ball above the ground before it was dropped.

c) Calculate the height the ball rebounds to.

d) Why is your answer to **c)** smaller than the answer to **b)**?

e) At what time does the ball hit the ground? Justify your answer.

11 A block of a material of mass 1 kg is heated using a 50 W electric heater. The heater is switched on for five minutes and the temperature of the block increases from 20 °C to 49 °C.

a) Calculate the specific heat capacity of the block.

b) The specific heat capacities of some substances are shown.

Material	Specific heat capacity in J/kg °C
Aluminium	902
Copper	386
Steel	500
Lead	128

Using the table, identify the material used to make the block.

12 An electric kettle rated at 2000 W is used to heat 0.50 kg of water. The water, at an initial temperature of 20 °C, is heated to its boiling point.

a) What is the shortest time it will take to boil the water?

b) When the time to boil the water was measured with a stopwatch, its value was found to be greater than the value calculated in **a)**.

Explain the difference in the times.

c) The kettle does not automatically switch off when the water boils due to a faulty thermostat. After the water has reached its boiling point, the kettle is left on for a further 100 s. Calculate the maximum mass of water which has been changed into steam during this time.

13 A hot water tank containing 210 kg of water is heated by an immersion heater. The initial temperature of the water is 15 °C. The immersion heater is connected to the 230 V mains supply and then switched on. The current in the element of the heater is 30 A. The heater is switched on for 24 minutes.

a) Calculate the energy supplied by the heater.

b) Calculate the final temperature of the water.

c) In practice the water does not reach this temperature. Suggest **two** reasons why?

14 A circuit containing a transformer and a motor is set up as shown.

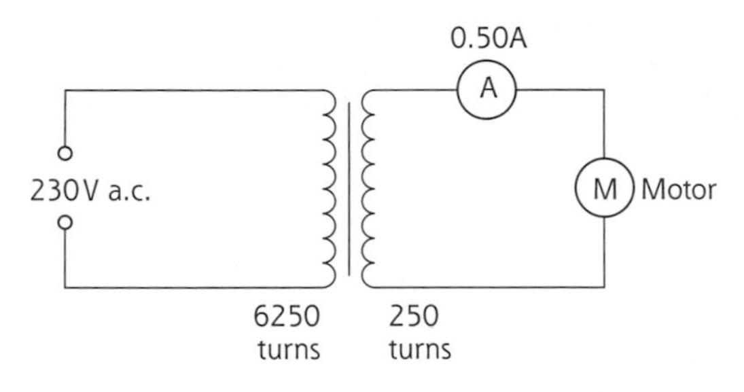

The transformer can be assumed to be 100% efficient.

The motor is used to lift a box at constant speed. The mass of the box is 0.3 kg. The motor lifts the box through a height of 0.9 m in 15 s. When the box is being lifted, the current in the secondary coil of the transformer is 0.50 A.

a) When the box is being lifted, calculate the current in the primary coil of the transformer.

b) How much electrical energy is supplied to the motor in 15 s?

c) Calculate the gain in gravitational potential energy of the box.

d) Find the efficiency of the motor during the lifting of the box.

Chapter 3

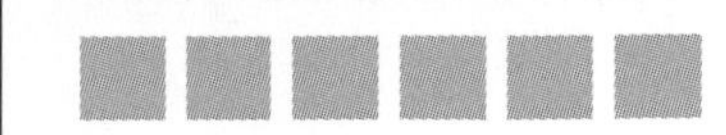

WAVES AND OPTICS

3.1 Waves

- A wave transfers energy.
- Radio, television signals and light are transmitted through air at a speed of 300 000 000 m/s.
- Amplitude of a wave is the height of the wave measured from the centre of the wave to the top of the crest or to the bottom of the trough.
- Frequency of a wave is the number of waves per second and is measured in hertz.
- The wavelength of a wave is the distance from one point of the wave until the pattern repeats itself. It is measured in metres.
- To calculate the speed of a wave use $v = f\lambda$ or $v = \frac{d}{t}$ where v is the speed of the wave, f is the frequency, λ is the wavelength, d is the distancc travelled, t is the time taken.
- In a given material a wave travels at constant speed.
- In a transverse wave the particles of the material the wave is travelling through vibrate at right angles to the direction the wave is travelling. An example of this is light waves.
- In a longitudinal wave the particles of the material the wave is travelling through vibrate parallel to the direction the wave is travelling. An example of this is sound waves.
- The members of the electromagnetic spectrum, from short to long wavelengths, are in this order: gamma rays, X-rays, ultraviolet, visible light, infrared, microwaves, TV and radio waves.

QUESTIONS

Example question

A boat flashes a light to the shore at the same time as it sounds a horn. A walker on the shore sees the flash of light and 6 s later hears the sound of the horn. How far is the boat from the shore?

Solution

Using the speed of sound in air from the data sheet on p. 98 as 340 m/s.

$d = vt$

$= 340 \times 6$

$= 2040$ m

1 During a thunderstorm, Tony hears the thunder 5 s after he sees the lightning. He estimates that the storm is 1700 m away.

a) What does Tony know about the speed of light compared to the speed of sound?

b) Calculate the speed of sound in air.

c) A short time later Tony hears the thunder 10 s after he sees the lightning. What does this tell Tony about the storm?

2 Two groups of pupils are attempting to measure the speed of sound. One group has an electrical circuit, which has a lamp and horn which can be switched on at the same time. The second group has stopwatches. The groups stand 500 m apart. The second group start their stopwatches when they see the light.

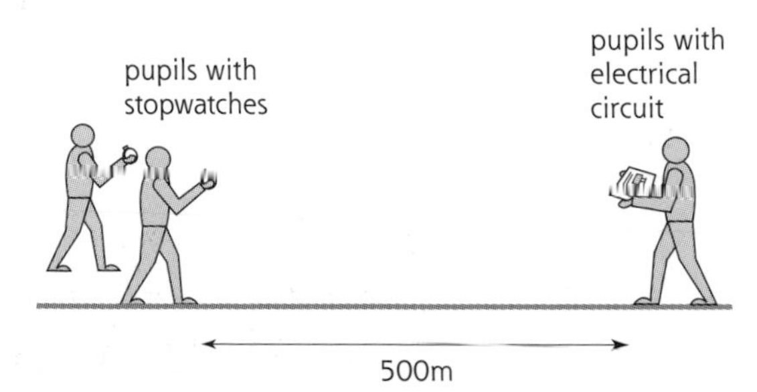

a) When do the second group stop their stopwatches?

b) The average time recorded on the stopwatches is 1.52 s. Calculate the value of the speed of sound in air obtained by the pupils.

c) A pupil suggests that the distance should be halved to make the measurements easier. Explain why this idea will make the estimate of the speed of sound less accurate.

3 An aircraft comes into land and passengers in the terminal building see the aircraft. The sound of the engines is heard 7.5 s after the aircraft is first seen. The speed of sound in air is 340 m/s. Calculate how far away the aircraft is when the passengers first see it.

4 In the table shown, calculate the value of each missing quantity.

	Distance (m)	Time (s)	Speed (m/s)
a)	6750	4.5	
b)	1980	6.0	
c)		0.002	1550
d)		6.7	330
e)	4290		780
f)	102		340

5 Fred is standing facing a hill. He shouts towards the side of the hill and hears the echo 6.4 s later. How far away is the side of the hill?

6 A fishing boat is trying to find a shoal of fish.

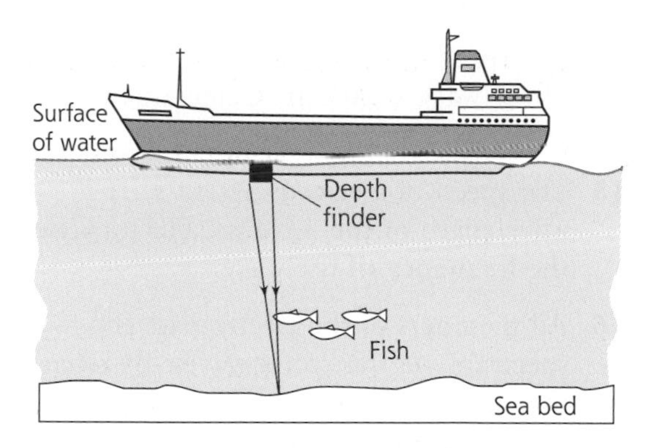

a) The depth finder emits sound waves. The sound waves travel at a speed of 1400 m/s in water. When the finder emits a sound an echo is heard 1.8 s later. How deep is the sea at this point?

b) The echo time suddenly changes to 1.0 s. How far is the shoal of the fish from the seabed?

c) The sound signal emitted by the depth finder can be displayed on the screen of an oscilloscope. The reflected signal has smaller amplitude than the transmitted signal. Explain why this occurs.

7 High frequency sound waves are used to check for flaws in a piece of metal. A combined transmitter and detector unit is used to send the sound waves into the metal. When the sound comes across a flaw, the sound waves are reflected back to the detector. In one test the sound is detected 0.8 ms after it was transmitted. The speed of sound in the metal is 5000 m/s. Calculate the distance of the flaw from the unit.

8 Tidal waves carry wood and other materials to the beach. State the quantity transferred by any type of wave.

9 For each of the waves shown in the diagrams below calculate:

i) the amplitude of the wave
ii) the wavelength of the wave.

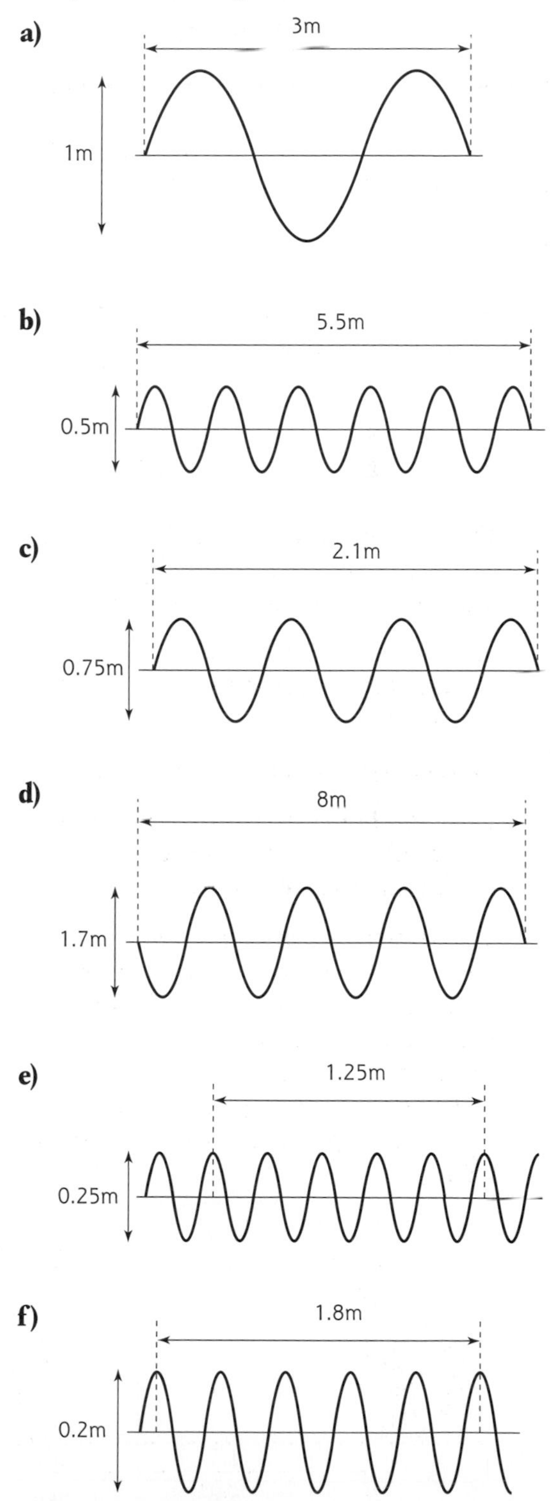

10 A wave has a frequency of 50 Hz and a wavelength 0.3 m. Explain what is meant by:

a) a frequency of 50 Hz

b) a wavelength of 0.3 m.

11 For each of the following calculate the frequency of the waves.

a) 25 waves pass a point in 10 s.

b) 76 waves pass a point in 9 s.

c) 136 waves pass a point in 19.5 s.

d) 6000 waves pass a point in 25 s.

e) 12000 waves pass a point in five minutes.

f) 7.5×10^3 waves pass a point in 6 ms.

g) 9.4×10^6 waves pass a point in 4 s.

h) 8 waves pass a point in 2 ms.

Example question

Sound waves have a frequency of 16 kHz and a wavelength of 0.02 m.

Calculate the speed of the waves.

Solution

$f = 16\,000$ Hz
$\lambda = 0.02$ m
$v = f\lambda$
$= 16\,000 \times 0.02$
$= 320$ m/s

12 In the table shown, calculate the value of each missing quantity.

	Frequency (Hz)	Wavelength (m)	Speed (m/s)
a)	4.0	7.7	
b)	20	1.8	
c)	200		300
d)	1.5		120
e)		1.5	200
f)		3.0	750

13 The wavelength of a wave is 1.2 m. The frequency of the wave is 25 Hz. Calculate the speed of the wave.

14 The speed of a wave is 12 m/s. The frequency of the wave is 20 kHz. Calculate the wavelength of the wave.

15 The speed of a wave is 360 m/s. The wavelength of the wave is 0.003 m. Calculate the frequency of the wave.

16 All members of the electromagnetic spectrum are transverse waves. By referring to the energy transfer explain what is meant by a transverse wave.

17 What is meant by the statement that sound waves are longitudinal?

18 State the speed at which all electromagnetic waves travel.

19 In the table shown, calculate the value of each missing quantity.

	Frequency	Wavelength (m)	Speed (m/s)
a)	4.2 MHz	75	
b)	35 kHz	4.3×10^{-2}	
c)	6×10^{14} Hz		1.25×10^8
d)		5×10^{-7}	3.0×10^8
e)		7.5×10^{-7}	2.0×10^8
f)	5 MHz		2200

20 The table shows the electromagnetic spectrum in order of increasing wavelength. Three parts labelled A, B and C of the electromagnetic spectrum are missing. Name the three radiations labelled A, B and C.

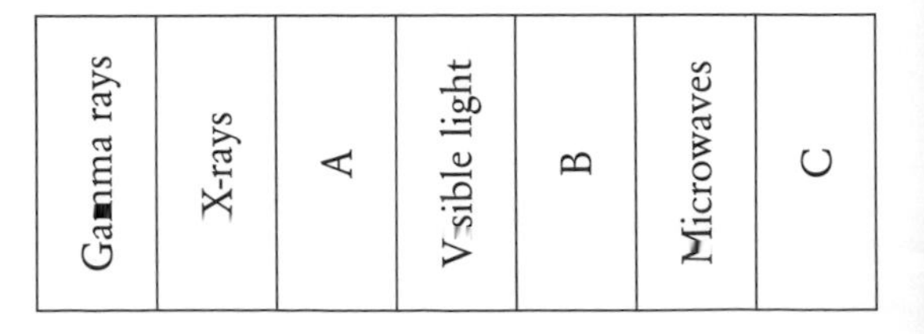

21 One type of mobile phone operates using signals with a frequency of 9.0 GHz. To detect the signal the length of the receiving aerial is equal in size to half a wavelength of the transmitted signal. Calculate the length of the receiving aerial.

22 State the detector used for each of the following radiations:

a) X-rays

b) ultraviolet

c) microwaves.

23 Thermograms are used to detect heat radiation from a person. This can be used to check if a drug is working.

a) What is the name given to the part of the electromagnetic spectrum which the thermogram detects?

b) How does the wavelength of this radiation compare with that of visible light?

3.2 Reflection

- An optical fibre is a thin piece of flexible glass along which light can be transmitted by total internal reflection.
- Modern television and telephone systems use electrical cables and optical fibres.
- When light is reflected from a plane mirror the angle of incidence equals the angle of reflection.
- A curved reflector brings the parallel rays of received signals to a focus, but reflects transmitted signals coming from the focus of the reflector into a parallel beam.
- Curved reflectors are used in satellite communication.
- Total internal reflection occurs when a ray of light does not leave the face of a piece of glass but reflects from that face back into the glass.
- The critical angle is the angle of incidence inside the glass block, which causes the ray of light to pass along the surface of the glass when it emerges.
- An optical fibre transmission system changes electrical signals into light, sends the light along a fibre system and then changes the light back to electrical signals at the receiving system.

QUESTIONS

1 The diagrams shows three plane mirrors.

a) What name is given to the dotted line in the diagrams?

b) Copy and complete the diagrams to show what happens to each incident ray.

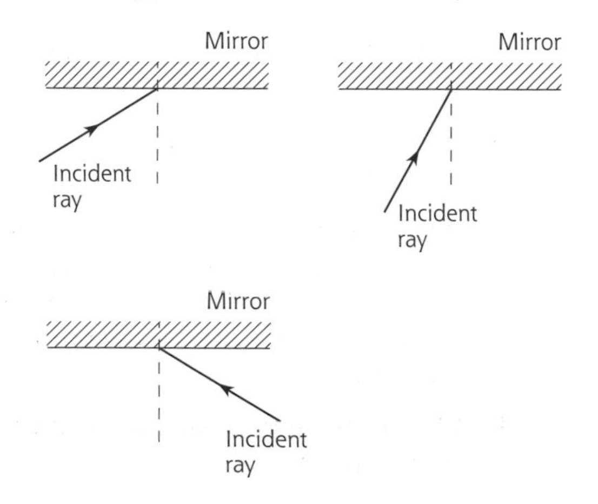

2 The diagram shows part of an optical fibre.

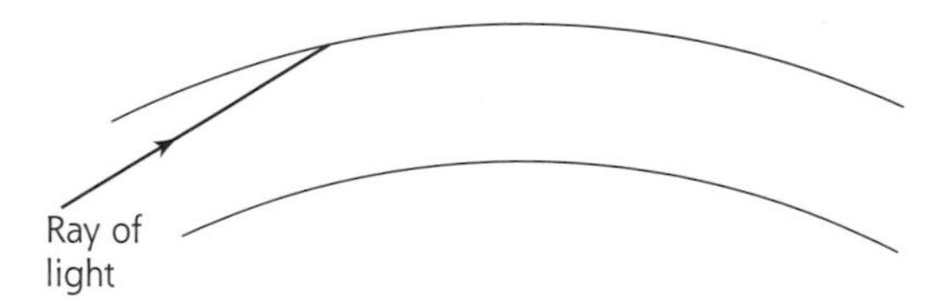

a) Show the passage of a ray of light along the optical fibre.

b) The optical fibre is 45 km long. Calculate the time for the light signal to be sent from one end of the fibre to the other end.

c) There is a suspected fault in the fibre. The signal returns to the source after being reflected from the fault. The time between the transmitted and returned signal is 0.22 ms. How far along the fibre is the fault?

3 The diagram shows a ray of light incident at point P on a mirror.

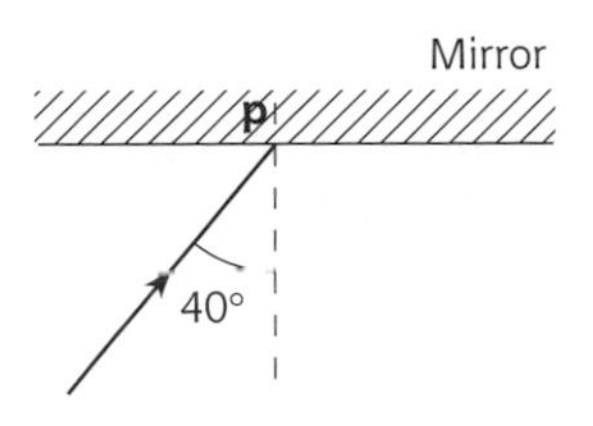

a) Copy and complete the diagram to show what happens to the ray of light.

b) Label your diagram to show the angle of incidence, angle of reflection and the normal.

c) What is the value of the angle of reflection?

4 Signals from a transmitter are sent out to a curved reflector.

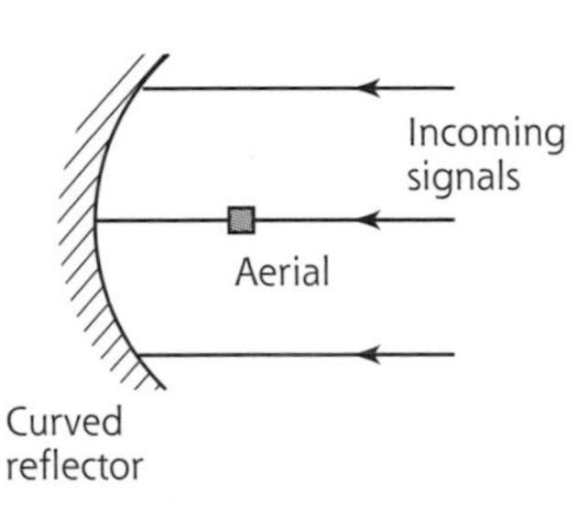

a) Copy the diagram to show the effect the curved reflector has on the incoming signals.

b) The curved reflector is replaced by a reflector which is more curved.

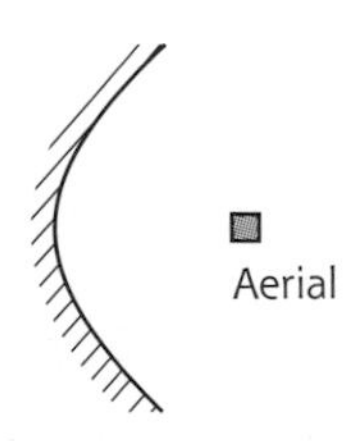

What effect, if any, will this have on the signals received at the aerial?

5 The diagram shows the path of a ray of red light in a semi circular glass block. The critical angle for the light in the glass is 42°.

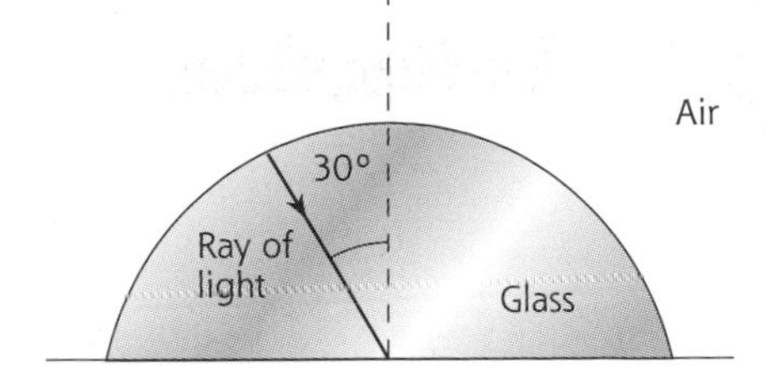

a) Complete the diagram to show the path of the ray of light as it passes out of the block.

b) Describe what happens if the angle of incidence is 45°.

6 The diagram shows the path of a ray of blue light incident on a point P of the glass prism. The critical angle for this light in the prism is 42°.

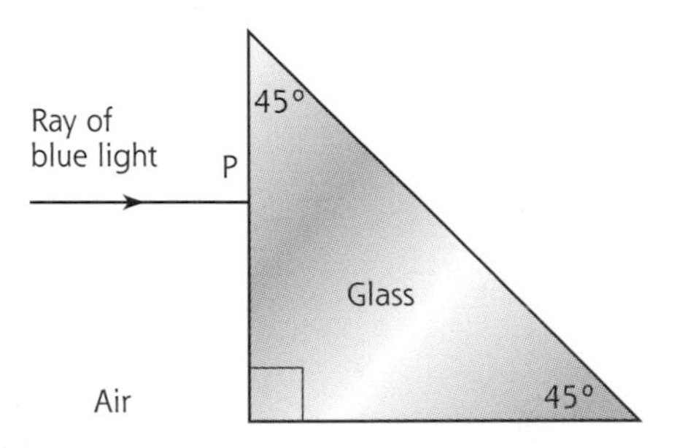

a) Explain what is meant by critical angle.

b) Copy and complete the diagram to show the passage of the ray of light through the prism.

7 Communication systems use two different ways to transmit information. One of these is an optical fibre system.

a) State the name of the other type of system.

b) What type of signal is sent along an optical fibre?

c) When telephone signals are sent down an optical fibre, state the energy changes that take place from the transmission to the reception of the signal. Start with electrical.

d) What is the main advantage of using optical fibres?

3.3 Refraction

- Refraction of light occurs when light travels from one material to another. It is caused by the speed of light changing.
- When a ray of light travels from air to glass the ray bends towards the normal and the angle of incidence is greater than the angle of refraction.
- A converging lens brings parallel rays of light to a focus but a diverging lens spreads the rays out.
- Ray diagrams are used to locate the position of an image formed by a lens.
  - One ray is drawn from the top of the object parallel to the central line called the principal axis. This ray then passes through the principal focus of the lens.
  - A second ray passes straight from the top of the object through the centre of the lens without changing direction.
  - Where the rays intersect is the top of the image formed.
- The image can be described in three ways:
  - It is real or virtual. A real image can actually be seen on a screen. A virtual one is one that appears to come from that point.
  - It is magnified or diminished.
  - It is upright or inverted.
- For a converging lens, the image of an object placed at a distance of:
  - more than two focal lengths is real, diminished and inverted.
  - between one and two focal lengths is real, magnified and inverted.
  - less than one focal length is virtual, magnified and upright.
- The power of a lens in dioptres is calculated as:

$$\text{Power} = \frac{1}{\text{focal length in metres}}$$

- For a converging (convex) lens the power is positive but for a diverging (concave) lens the power is negative.
- Long sight is when distant objects are seen clearly but near objects are blurred. A converging lens will correct the problem.
- Short sight is when distant objects are blurred but near objects are clear. A diverging lens will correct the problem.

QUESTIONS

1 The diagram shows a ray of red light incident at point P on a glass block.

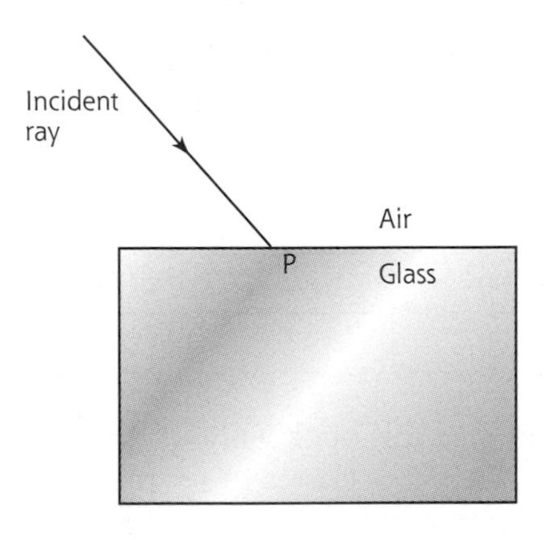

a) A pupil writes that this ray of light is refracted. What is meant by refracted?

b) Copy the diagram to show what happens to the ray of light. Label the angle of incidence, the angle of refraction and the normal.

c) The angle of incidence in air is 40°. State a possible value for the angle of refraction.

2 The diagrams show two types of lens, X and Y.

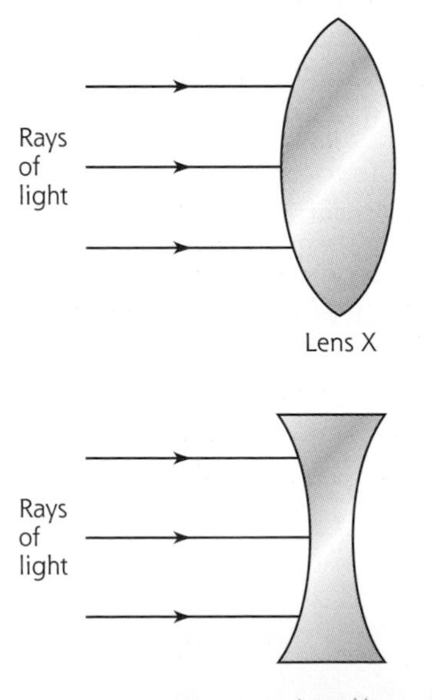

a) Complete the diagrams and show what happens to the rays of light.

b) The focal length of lens X is 100 mm. Explain what is meant by this statement.

3 A convex lens is used in an overhead projector. The diagram below shows an object which is placed at a distance of 220 mm in front of the lens. The focal length of the lens is 120 mm.

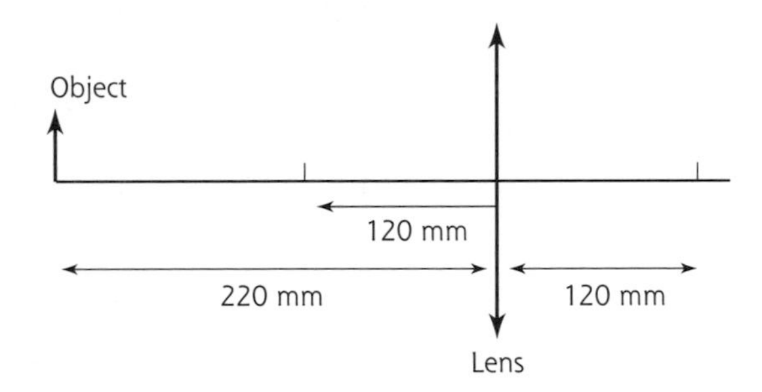

a) Draw a suitable diagram to scale to show how the final image is produced by the lens.

b) State whether the image is real or virtual and if it is magnified or diminished.

4 An insect is viewed using a magnifying glass.

The magnifying glass uses a convex lens of focal length 40 mm. The insect is placed 25 mm from the centre of the lens.

a) Draw a diagram to show how the image of the insect is produced. Use an arrow to represent the insect. Make the height of the arrow 20 mm.

b) What is the size of the magnified image?

5 A slide is placed 80 mm from the centre of a convex lens in a slide projector. The focal length of the lens is 60 mm. Draw a diagram to show how the image of the slide is produced. Use an arrow to represent the slide.

6 Different objects are placed in front of different convex lenses. Each situation is described in **a)**, **b)** and **c)**.

a) An object is placed 120 mm in front of a convex lens. The focal length of the lens is 50 mm.

b) An object is placed 80 mm in front of a convex lens. The focal length of the lens is 60 mm.

c) An object is placed 20 mm in front of a convex lens. The focal length of the lens is 30 mm.

For each situation **a)**, **b)** and **c)**
i) Draw an arrow 20 mm long to represent the object.
ii) Draw a ray diagram to show how the image of the object is formed.
iii) Measure and state the distance from the lens to the position of the image.
iv) State whether the image is real or virtual.
v) State whether the image is magnified or diminished compared to the object.

Example question

A converging lens has a focal length of 100 mm. Calculate the power of the lens.

Solution

The focal length must be changed to metres.
Focal length is 0.1 m.

$$\text{Power} = \frac{1}{0.1} = 10\,\text{D}$$

If a diverging lens is used then the power would be negative.

7 A patient is given a prescription describing the correcting lens required for his spectacles. The prescription states that a lens of + 2.5 D is required.

a) What type of sight defect does the person have? Explain your answer.

b) Calculate the focal length of the lens for the spectacles.

8 In the table shown, calculate the value of each missing quantity. State the type of any missing lens.

	Focal length (mm)	Power (D)	Type of lens
a)	150		Convex
b)	200		Concave
c)		2.5	
d)		−4	
e)	500		Concave
f)	300		Convex

9 Bill can read this book without the aid of glasses. He needs glasses to see the number on an approaching bus.

a) Name the sight defect that Bill has.

b) What type of lens is needed to correct this defect?

c) The focal length of the lens in Bill's glasses is 0.6 m. Calculate the power of the lens.

10 A patient requires glasses. The prescription for the lenses is shown in the table.

Right eye	Left eye
Sph −1.2 D	Sph −1.6 D

a) What sight defect does the person have? Explain your answer.

b) Calculate the focal length of the lens used with the left eye.

c) Which eye requires a correcting lens with a greater focal length?

Exam-style questions

1 **a)** A glass slide is placed 300 mm from a converging lens. The focal length of the lens is 200 mm. Draw a suitable diagram to show how an image is produced by the lens.

b) **i)** State whether the image is real or virtual.
ii) Is the image magnified compared to the object?

2 A clear object is placed 150 mm from a converging lens as shown below. The image is produced 300 mm from the lens on the other side. The image is twice as large as the object. By completing the diagram calculate the focal length of the lens.

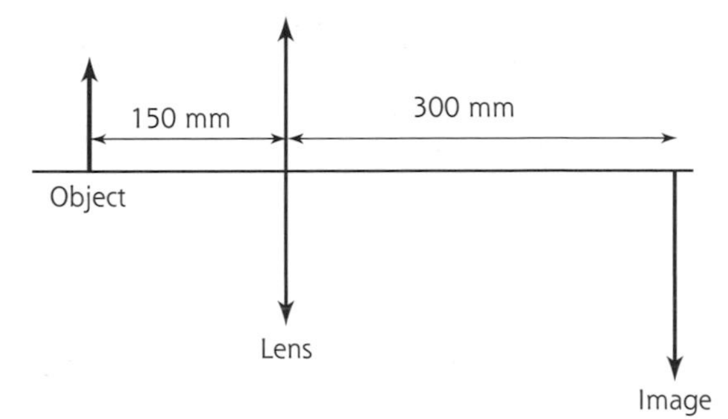

3 Lasers can be used to treat a wide range of conditions. Details of some lasers are given in the table below.

Type of laser	Wavelength (nm)	Power
Argon	490	1.0 W
Helium neon	633	5.0 mW
Gallium arsenide	910	1.0 mW
Neodymium YAG	1064	50 W
Carbon dioxide	10 600	15 W
Excimer	193	20 W

$1\ \text{nm} = 10^{-9}\ \text{m}$

The wavelength of light ranges from 400 nm to 700 nm.

a) Which lasers emit infra red radiation?

b) Calculate the frequency of the light emitted by the helium neon laser.

c) One laser emits radiation in the ultra violet radiation.
i) State which laser emits ultraviolet radiation.
ii) Name a detector for this type of radiation.

4 A variety of lenses are used to correct sight defects.

a) Tom has difficulty seeing distant objects but can read this page clearly.
i) Which sight defect is he experiencing?
ii) What kind of lens will correct this problem?

b) A lens has a power of −3.25 D. Calculate the focal length of this lens.

c) Describe an experiment to measure the focal length of a converging lens. In your description use a labelled diagram to show clearly the apparatus used and any measurements taken.

5 During a building accident, people are thought to be trapped below the rubble. They can be detected using heat radiation.

a) What is the name given to this type of radiation?

b) How does the wavelength of this radiation compare with that of visible light?

c) When people go to ski in mountains, less radiation is absorbed than at lower altitudes. This means that there is more ultraviolet radiation than normal.

What is the danger of excessive exposure to ultraviolet radiation?

6 A ray of light passes from air into glass.

a) **i)** What name is given to this effect?
ii) Explain how this effect happens.

b) i) The diagram shows the passage of a ray of red light until it meets the surface of the glass at P. The critical angle for this light in the glass is 42°. Copy and complete the diagram to show what happens now to the ray of light.

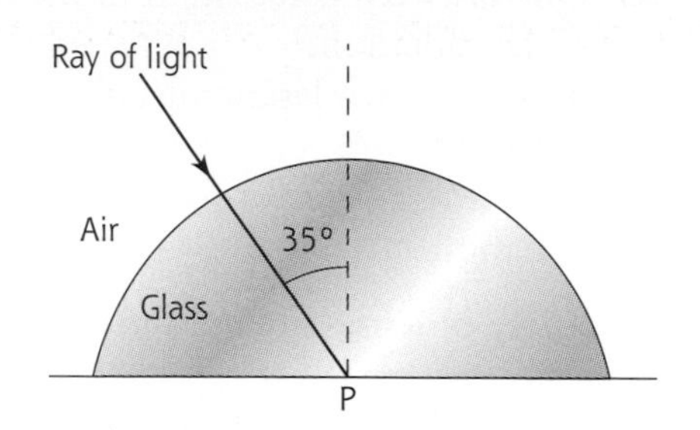

ii) Label your diagram to show the angle of incidence and the angle of refraction.
iii) Describe what would happen as the angle of incidence is increased.

Chapter 4

RADIOACTIVITY

4.1 Ionising radiations

- The structure of an atom includes three different types of particles namely protons (+), neutrons and electrons (−).
- Radiation energy may be absorbed in a substance.
- Radiation can kill or change the nature of living cells.
  - Alpha radiation is absorbed by paper and will travel about 5 cm in air.
  - Beta radiation is absorbed by a few millimetres of aluminium.
  - Gamma radiation is absorbed by a few centimetres of lead.
- An alpha particle is a helium nucleus; a beta particle is a fast-moving electron and gamma radiation is an electromagnetic radiation with a very short wavelength.
- Ionisation is the gain or loss of an electron to produce a charged particle.
- Alpha particles produce much greater ionisation density than beta particles or gamma rays.
- Ionisation is used in a detector of radiation.
- Radiation can destroy cells and is used in cancer treatment.
- Gamma cameras use radiation to examine the organs of the body.

QUESTIONS

1 A sheet of paper, a sheet of aluminium 5 mm thick and a sheet of lead 2 cm thick are placed in front of a Geiger Muller tube connected to a counter. A stream of alpha particles, beta particles and gamma rays are directed onto the sheet of paper.

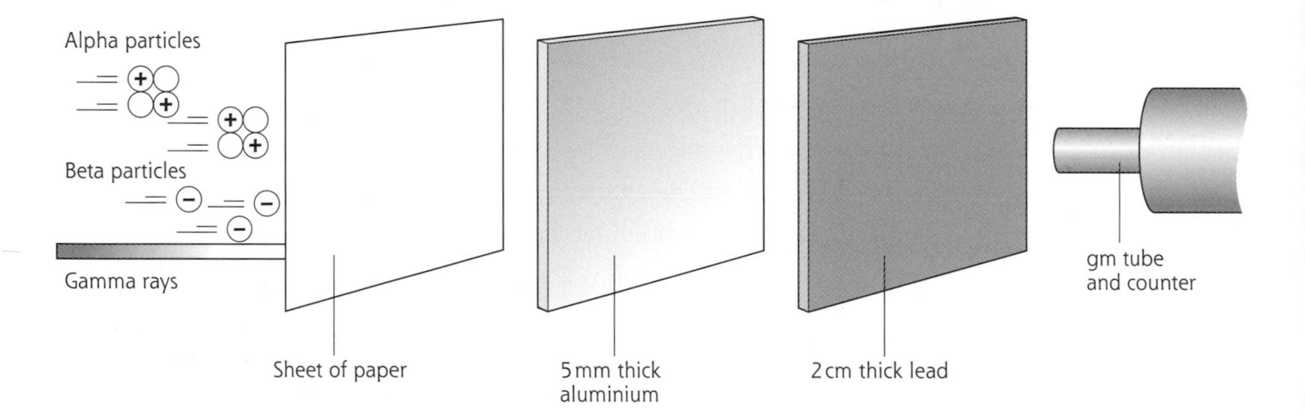

a) Which radiation/radiations passes/pass through the paper?

b) Which radiation/radiations passes/pass through the aluminium?

c) Which radiation/radiations is/are detected by the counter?

2 An atom is made up of three different types of particles. For each statement below state the name of the particle that is being described.

a) Has a mass of about 1/2000 of a proton and is not in the nucleus.

b) Has no charge.

c) Has a positive charge.

3 A student makes the following statements about three different types of radiation:

a) Produces a large amount of ionisation and has a positive charge.

b) An electromagnetic wave which has a short wavelength.

c) Absorbed by a few mm of aluminium and has a negative charge.

Identify which radiation is being described by each statement.

4 Nuclear radiation produces ionisation.

a) What is meant by ionisation?

b) A certain nuclear radiation can easily be absorbed by paper or skin. If swallowed, however, it produces a large amount of damage inside the body.
i) What is the name of this radiation?
ii) Why is this radiation so dangerous inside the body?

5 Photographic film is often used in radiation badge detectors. On what effect does this film depend in order to detect radiation?

6 A film badge is shown in the diagram below. State and explain which radiations will be detected by regions X and Y.

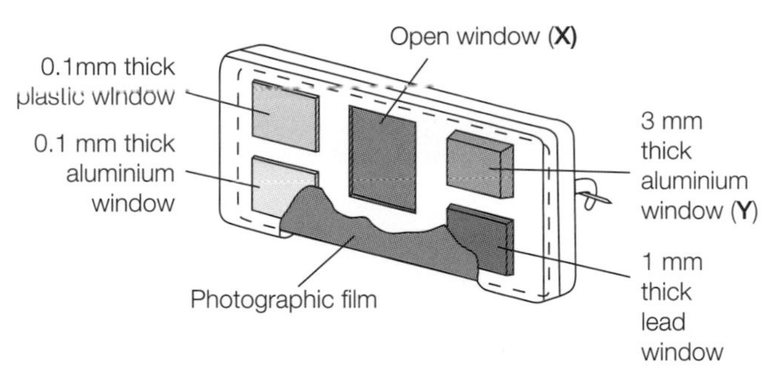

7 Radiotherapy involves using radiation to treat cancers. What does this treatment do to the cancerous cells of a tumour?

8 Other than the treatment of cancer, describe another use of gamma radiation in medicine.

9 Gamma radiation is used to detect whether an organ inside the body is working properly.

a) Why is gamma radiation used rather than alpha or beta radiation?

b) i) The diagram below shows a scan from the kidneys. State which kidney appears not to be working properly.
ii) Explain your choice.

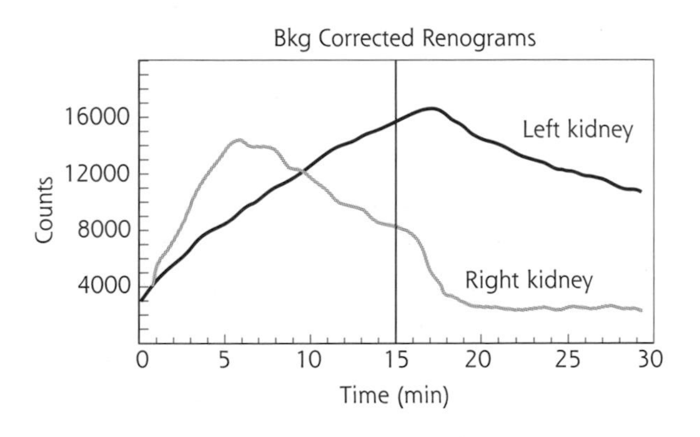

10 A smoke detector is shown below. When smoke passes between the radioactive source and the detector, the absorption of the radiation makes the detector sound an alarm.

a) Why is a source of alpha radiation used for the radioactive source?

b) i) Give an approximate size for the gap between the source and detector.
ii) Explain your choice.

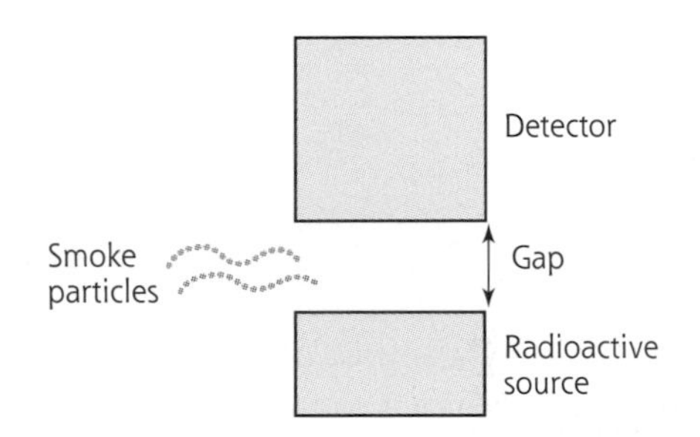

11 The thickness of cardboard packaging is measured using a thickness gauge. The gauge consists of a radioactive source and a detector as shown below.

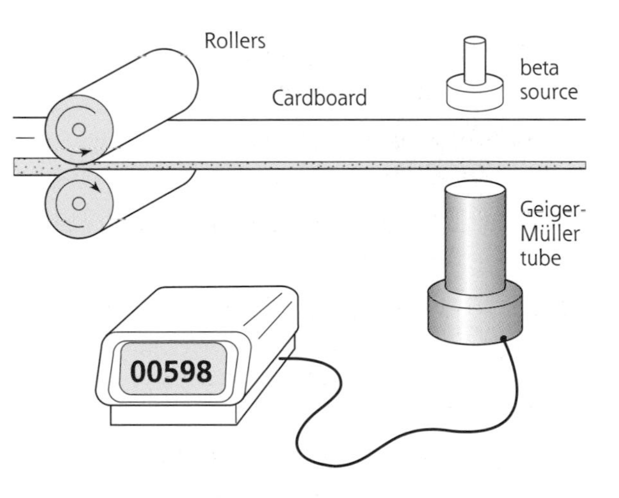

a) State and explain which source of radiation (alpha, beta or gamma) should be used.

b) Explain how this detector can be used to show that the thickness of the cardboard has changed.

c) A visitor to the factory thinks that any cardboard produced in that factory must be radioactive. Describe how you might convince the visitor that this will not be true.

12 A brain tumour can often be treated by placing a source of alpha radiation near the region of the tumour.

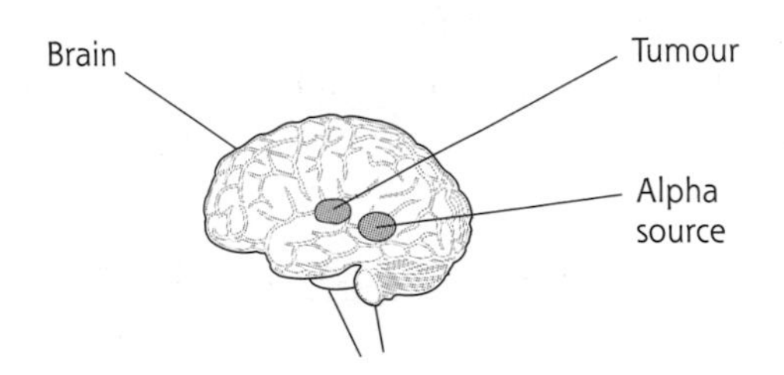

Explain why a source of alpha radiation is used rather than a source of beta or gamma radiation.

4.2 Dosimetry

- The activity of a radioactive source is measured in becquerels (Bq), where one becquerel is the decay of one nucleus in the radioactive source per second.
- $A = \frac{N}{t}$ where A is the activity of the source, N is the number of decays and t is the time in seconds.
- The absorbed dose D is the energy absorbed per kilogram of the absorbing material and is measured in grays (Gy), which are joules per kilogram.
- $D = \frac{E}{m}$ where E is the energy absorbed and m is the mass of tissue absorbing the radiation.
- The risk of biological harm from radiation depends on:
  - the absorbed dose
  - the kind of the radiation
  - the body organs or tissue exposed to the radiation.
- A radiation weighting factor w_R is given to each kind of radiation as a measure of its biological effect. These values are given in the data sheet.
- Equivalent dose $H = D\,w_R$ and is measured in sieverts (Sv).
- There are several sources of background radiation, which can be split into natural and man-made.

QUESTIONS

Example question

In a source of radiation 1.2×10^{11} nuclei decay in a time of two minutes. Calculate the activity of the source.

Solution

$A = \frac{N}{t}$

$N = 1.2 \times 10^{11}$

$t = 2$ minutes $= 120$ s

$A = \frac{1.2 \times 10^{11}}{120}$

$= 1 \times 10^{9}$ Bq

1 The activity of a radioactive source is stated to be 5 MBq.

a) What is meant by this statement?

b) After a period of time what happens to the activity of the source?

2 In the table shown, calculate the value of each missing quantity.

	Number of decaying nuclei	Time	Activity (Bq)
a)	5×10^6	250 s	
b)	70 000	1 min 40 s	
c)	3.5×10^6		2×10^6
d)	80 000 000		8×10^7
e)		300 s	7.2×10^6
f)		540 s	5 000

3 In a source of radiation 2400 nuclei decay in one minute. Calculate the activity of the source.

4 The activity of a radioactive source is 1.5 MBq. Calculate the number of radioactive nuclei in the source that decay in 15 minutes.

5 A radioactive source has an activity of 300 kBq. In the source 9×10^6 nuclei disintegrate during an experiment. Calculate the time that the source was used for during the experiment.

6 A source of radiation has an activity of 5 MBq and is used for 10 minutes to treat a tumour. Calculate the number of radioactive nuclei that disintegrate in this time.

Example question

Radiation is incident on a sample of tissue. The mass of tissue is 2.3 kg. The energy absorbed by the sample is 9.2×10^{-3} J.

Calculate the absorbed dose taken in by the tissue.

Solution

$$D = \frac{E}{m}$$

$$= \frac{9.2 \times 10^{-3}}{2.3}$$

$$= 4 \times 10^{-3}\ \text{Gy}$$

7 In the table shown, calculate the value of each missing quantity.

	Energy (J)	Mass (kg)	Absorbed dose (Gy)
a)	0.30	5.0	
b)	1×10^3	2.0×10^{-2}	
c)		10	5.0×10^{-3}
d)		50	8.0×10^{-6}
e)	4.0×10^{-4}		8×10^{-6}
f)	0.2		4×10^{-3}

8 A sample of tissue is exposed to 50 mGy of gamma radiation. The mass of the sample is 0.55 kg. Calculate the energy that the sample receives.

9 A sample of tissue absorbs 1.3 mJ of energy. The absorbed dose received by the tissue is 2 mGy. Calculate the mass of the tissue.

10 During treatment for cancer a tumour absorbs 2.88 mJ of energy from a source of gamma radiation. The mass of the tumour is 0.72 kg. Calculate the absorbed dose received by the tumour.

Example question

A worker testing equipment uses slow neutrons. The radiation weighting factor for the slow neutrons is 3. A sample of tissue receives an absorbed dose of 6 μGy. Calculate the equivalent dose received by the sample of tissue.

Solution

$D = 6 \times 10^{-6}$ Gy

$w_R = 3$

$H = D\,w_R$

$= 6 \times 10^{-6} \times 3$

$= 1.8 \times 10^{-5}$ Sv

In the following questions unless told otherwise you should refer to the data sheet on p. 98 for radiation weighting factors for different radiations.

11 State the three factors which determine the risk of biological harm from radiation.

12 In the table shown, calculate the value of each missing quantity

	Absorbed dose (Gy)	Radiation weighting factor	Equivalent dose (Sv)
a)	1×10^{-3}	1	
b)	5×10^{-6}	20	
c)		20	4.5×10^{-6}
d)		1	2.5×10^{-3}
e)	1×10^{-4}		2.0×10^{-3}
f)	1×10^{-4}		1.0×10^{-4}

13 A sample of tissue receives an absorbed dose of 0.05 Gy of fast neutrons. The mass of the sample is 0.5 kg. Calculate the equivalent dose received by the sample of tissue.

14 A sample of tissue is exposed to a single type of radiation. The equivalent dose received by the sample is 0.3 mSv. The absorbed dose received by the sample is 0.1 mGy. Identify the type of radiation the tissue is exposed to.

15 During exposure to fast neutrons, some tissue receives an equivalent dose of 200 μSv. Calculate the absorbed dose received by the tissue.

16 During some medical research a sample of tissue receives an absorbed dose of 10 mGy of radiation. The mass of the sample tissue is 700 g. Calculate the energy received by the tissue.

17 A tissue sample receives 25 μGy of gamma radiation during an investigation.

Calculate the equivalent dose received by the tissue.

18 A sample of tissue receives an absorbed dose of 10 μGy from alpha radiation and 25 μGy from gamma radiation. Calculate the total equivalent dose received by the sample.

19 In a test procedure, a sample of tissue is exposed to an absorbed dose of 0.2 mGy from fast neutrons and 10 μGy from beta radiation. Calculate the total equivalent dose received by the sample.

20 During radiation work in a reactor a tissue sample receives a total equivalent dose of 0.18 mSv. This consists of 120 μSv received from alpha radiation and an unknown equivalent dose received from fast neutrons.

a) Calculate the equivalent dose received from the fast neutrons.

b) Calculate the absorbed dose received from each type of radiation.

21 During a test a sample of tissue receives an equivalent dose of 450 μSv. The absorbed dose is 150 μGy. Calculate the radiation weighting factor and identify the type of radiation the tissue is exposed to.

22 Background radiation consists of natural and radiation produced by industry.

a) State two sources of natural radiation.

b) State two sources of industrial radiation.

4.3 Half-life and safety

- The activity of a radioactive source decreases with time.
- The half-life of a radioactive source is the time taken for half the radioactive nuclei to disintegrate.
- The safety precautions necessary when handling radioactive sources include handling the source with tongs, pointing the source away from you and not eating when working with a source.
- The equivalent dose received by a sample of tissue is reduced by shielding, by limiting the time of exposure or by increasing the distance from a source.
- The radioactive hazard sign is an international one and it should be displayed where sources are stored or are in daily use.

QUESTIONS

Example question

A radioactive source has a half-life of 24 days. The initial activity of the source is 2000 Bq. What is the activity of the source after 120 days?

Solution

Number of half lives $= \frac{120}{24} = 5$

Activity after one half life = 1000 Bq
Activity after two half lives = 500 Bq
Activity after three half lives = 250 Bq
Activity after four half lives = 125 Bq
Activity after five half lives = 62.5 Bq

1 Technetium 99 m is a radioactive source which is used in medicine. It has a half-life of six hours. Explain what is meant by a half-life of six hours.

2 A radioactive sample has a half-life of 20 minutes. The initial activity of the source is 2200 Bq. Calculate its activity after one hour.

3 The graph below shows how the activity of a radioactive source varies with time. Calculate the half-life of the source.

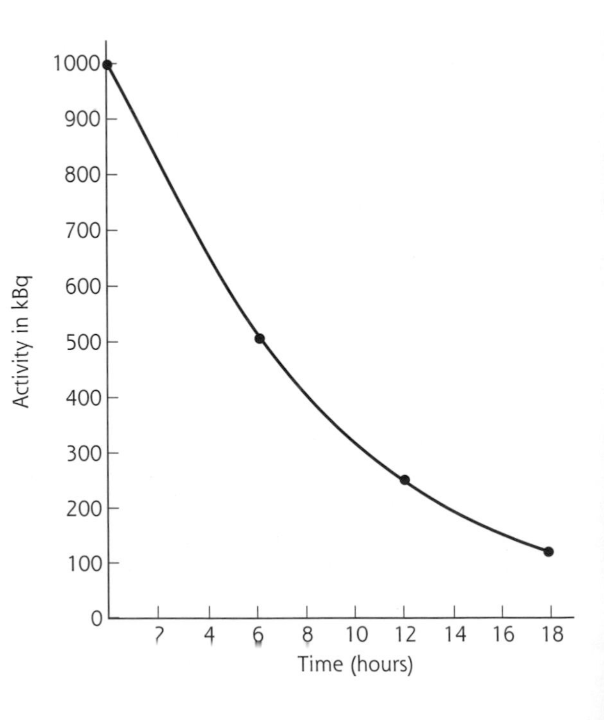

4 A radioactive source is stored in a lead-lined container.

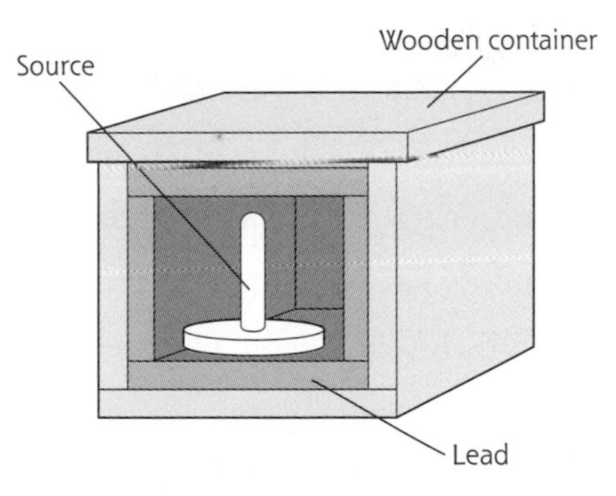

The count rate from the source is measured every four hours. The results are displayed in the table shown. The count rate is a measure of the activity of the source.

Time (hours)	Count rate from source (counts per minute)
0	168
4	125
8	84
12	60
16	42
20	30
24	21
28	15

a) Why is the source stored in a lead-lined container?

b) Calculate the half-life of the radioactive source.

5 After a nuclear reprocessing procedure a radioactive source is stored safely. The initial activity of the source is 800 MBq. The half-life of the source is five years. Calculate the activity of the source after 20 years.

6 A radioactive source is used in medical treatment. The initial activity of the source is 5 MBq.

The half-life of the source is eight days. Calculate the activity of the source after 40 days.

7 a) State the unit of activity of a radioactive source.

b) What happens to the activity of the source as time increases?

8 When prepared the initial activity of a source is 10 MBq. After 24 hours the activity is 625 kBq. Calculate the half-life of this source.

9 The half-life of radon-222 is 3.8 days. The initial activity of a sample of radon-222 is 500 kBq. Calculate the activity after 19 days.

10 Carbon-14 is used to date old samples of paper. The half-life of carbon-14 is 5730 years. The mass of a sample of carbon-14 is 70 mg. Calculate the mass of carbon–14 present after 17 190 years.

11 The half-life of cobalt-60 is 5.3 years. After 15.9 years, the activity of a sample of cobalt-60 is 60 kBq. Calculate the initial activity of the sample.

12 A sample of gold-198 has a mass of 200 g. After 10.8 days the mass of gold-198 in the sample is 12.5 g. Calculate the half-life of gold-198.

13 The initial activity of a radioactive source is 5 MBq. The half-life of the source is eight days. Calculate the time taken for the activity to fall to 625 kBq.

14 During a laboratory experiment, background radiation is measured to be 50 counts per minute (c.p.m.). A radioactive source is placed in front of a Geiger Muller tube which is connected to a counter. The total count rate is recorded at five minute intervals.

Time (minutes)	Recorded count rate (c.p.m.)	Source count rate (c.p.m.)
0	242	
5	202	
10	170	
15	146	
20	126	
25	110	
30	98	

The source count rate is a measure of the activity of the source.

a) Complete the last column of the table.

b) Draw a graph of source count rate against time. Use the graph to calculate the half-life of the radioactive source.

15 A radioactive source is placed in front of a detector, which is connected to a counter. The recorded count rate is 195 counts per minute (c.p.m.). The experiment is repeated 24 hours later and the recorded count rate is now 60 c.p.m. The background count rate on both occasions is 15 c.p.m. The source count rate is a measure of the activity of the source. Calculate the half-life of the source.

16 A radioactive source is placed in front of a detector connected to a counter. The count rate is measured as 422 c.p.m. The measurement is repeated 32 days later. The count rate is now 45 c.p.m. The average background count rate is 22 c.p.m.

a) Estimate the half life of the radioactive source.

b) Explain why this is an estimate of the half-life of the source.

c) Why was an average value for the background count rate taken?

17 A radioactive source is to be used in the examination of a kidney. The procedure involves taking measurements for 30 minutes. There are three radioactive sources, X, Y and Z, available.

Source X has a half-life of 10 minutes.

Source Y has a half-life of 6 hours.

Source Z has a half-life of 8 days.

a) Which source should be used?

b) Explain your answer.

18 State two safety precautions when handling radioactive sources.

4.4 Nuclear reactors

- There are advantages and disadvantages of using nuclear power for the generation of electricity.
- The advantages are that uranium is plentiful and no greenhouse gases are emitted.
- The disadvantages are that radiation is dangerous to humans and the disposal of radioactive waste involves special storage for a long time.
- Fission occurs when certain nuclei are bombarded by neutrons and become unstable. The nuclei break into smaller pieces and release neutrons and energy.
- A chain reaction occurs when neutrons released by fission go on to strike more nuclei and cause further fission reactions.
- A nuclear reactor has fuel rods, a moderator, control rods, coolant and a containment vessel.
- The disposal and storage of nuclear waste are difficult due to the long half-life of some sources and the long-term storage that is required.

QUESTIONS

1 Nuclear power is considered dangerous and unnecessary by some people. What arguments would you use to convince them that it is useful to generate power from nuclear reactions?

2 Your friend gives other views against using nuclear power. What arguments will he or she use?

3 Inside the nuclear reactor of a power station, uranium-235 nuclei are bombarded by neutrons. When a uranium-235 nucleus is absorbed by a neutron, the nucleus becomes unstable.

a) What name is given to the splitting of a uranium nucleus?

b) Explain what happens after the nucleus becomes unstable.

c) What type of energy is released in this process?

4 The diagram below shows a chain reaction.

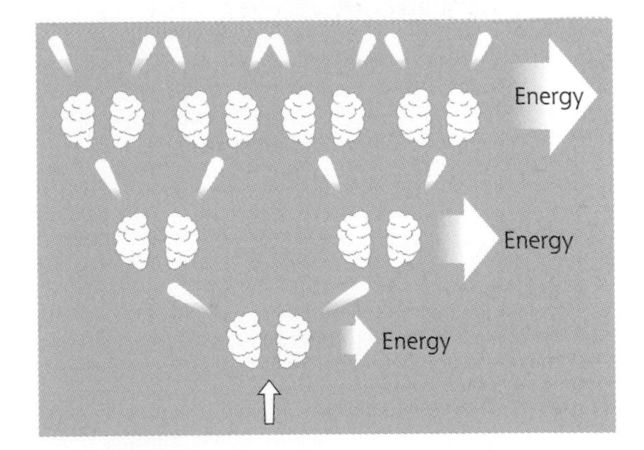

a) Explain what is meant by a chain reaction.

b) Why is it important that other neutrons are released in this process?

5 There are fuel rods in a nuclear reactor.

a) What is the purpose of the fuel rods?

b) Why do the fuel rods have to be replaced every few years?

c) State two reasons why there are problems in transporting used fuel rods from a nuclear power station to another site.

6 A nuclear power station contains a containment vessel. What is the function of the containment vessel?

7 The diagram below shows the key parts of a nuclear reactor.

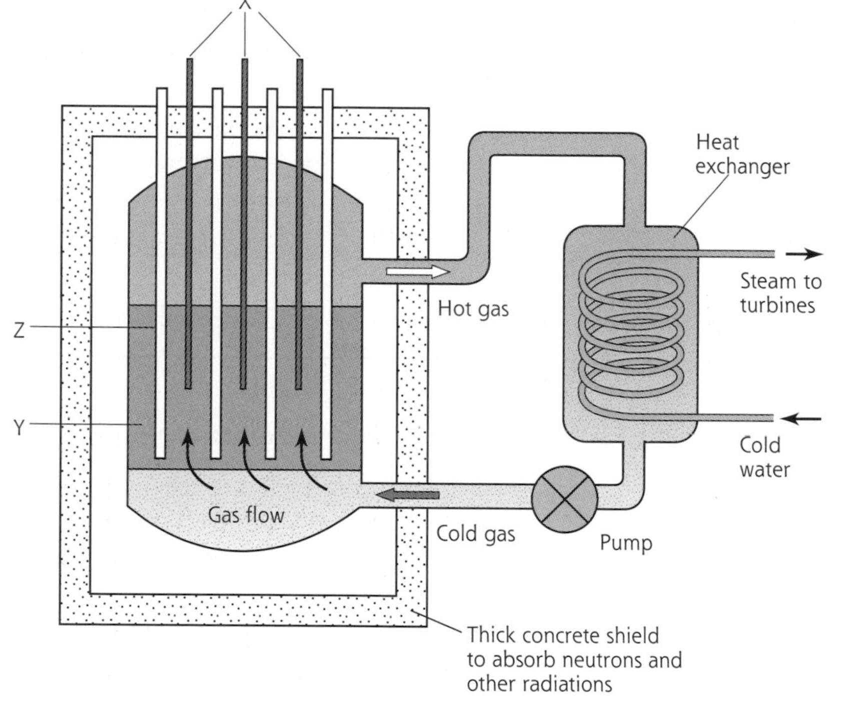

a) Name the parts labelled X, Y and Z.

b) Describe what each of these parts do in a nuclear reactor.

8 Two sites, X and Y, have been proposed which can be used to store nuclear waste.

Site X is in the countryside and is close to some hills and near the sea.

Site Y is close to a large city and has marshy ground nearby.

a) Explain why site X might be suitable for the storage of nuclear waste.

b) Explain why site Y might not be suitable for the storage of nuclear waste.

Exam-style questions

1 The diagram shows two types of radiation X and Y passing through different materials.

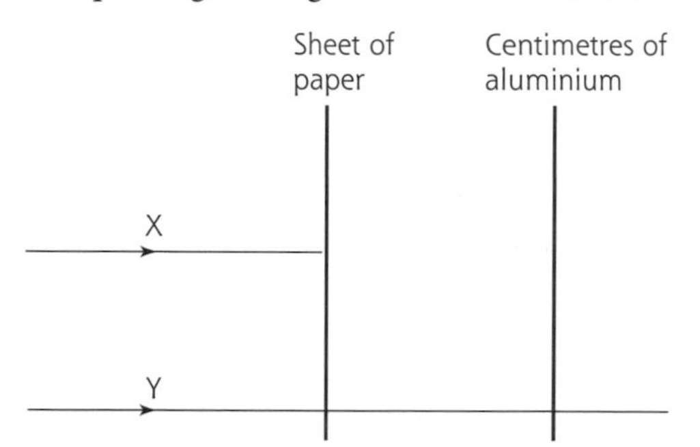

a) State the names of the two radiations labelled X and Y.

b) One of the radiations produces a high ionisation density.
i) What is meant by ionisation?
ii) Which radiation, X or Y, produces a higher ionisation density?

2 Technetium-99 m is a source of gamma radiation. The half-life of technetium-99 m is six hours.

a) Explain what is meant by a half-life of six hours.

b) A sample of technetium-99 m is combined with a drug and put into the body to go to a specific organ such as the kidneys. The gamma radiation emitted is detected by a gamma camera. Why is a source of gamma radiation used?

c) To treat a tumour with radiation, the radiation is directed at the tumour from different directions around the body as shown in the diagram. Explain why this method is used.

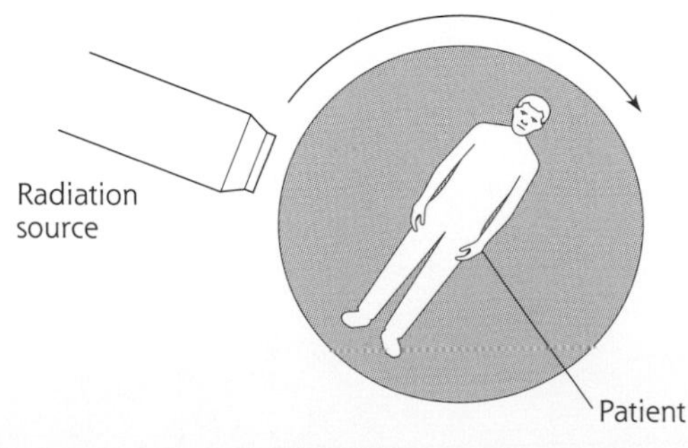

3 A radioactive source is used in a hospital. The half-life of the source is eight hours. The source emits gamma radiation.

a) On a Monday morning at 8a.m. the activity of the source is 16 kBq.

Calculate the activity of the source at 4p.m. on Tuesday afternoon.

b) **i)** An organ of the body is exposed to this radiation. The mass of the organ is 0.6 kg. The organ receives 0.03 J of energy. Calculate the absorbed dose received by the organ.
ii) Calculate the equivalent dose received by the organ.

c) The radioactive source has to be used in different parts of the hospital. It is transported in a container. What type of material should be used to construct the container?

4 A nuclear reactor has an output power of 480 MW. The reactor has 60 fuel rods. Each fuel rod produces 20 MW of power.

a) Calculate the efficiency of the reactor.

b) The process taking place in the fuel rods involves neutrons striking heavy nuclei and emitting further particles. State the name of this process.

c) State any precautions that should be taken when transporting the fuel rods to a reprocessing centre.

d) At the reprocessing centre the rods are then stored under water for a year. Explain why this is done.

ANSWERS

Chapter 1 ELECTRICITY AND ELECTRONICS

1.1 Circuits

1 A = electron, B = neutron, C = proton, D = nucleus

2

Conductor	Insulator
aluminium foil	glass
coin (2p)	air
iron nail	paper
metal ruler	plastic pen
	wooden pencil

3 a) 120 C, b) 2 C, c) 50 A, d) 0.2 A, e) 50 s, f) 220 s

4 a) 100 C, b) 360 C, c) 36 C

5 600 C; **6** 80 s; **7** 4 A

8 A = electrons, B = current, C = amperes, D = energy, E = volts

9 5 joules of energy are given to each coulomb of charge by the supply

10

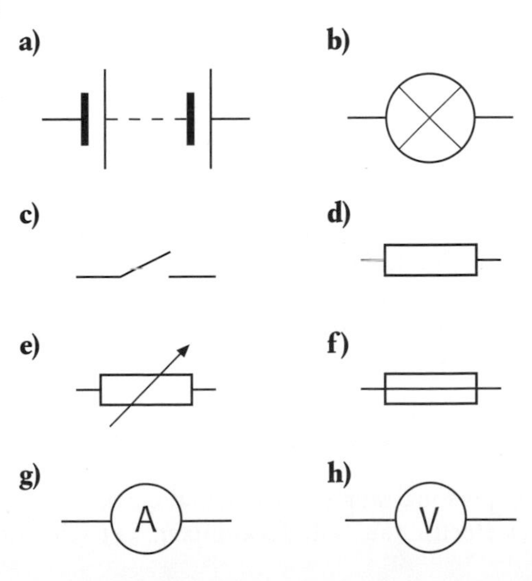

11 P = Ammeter, Q = Voltmeter

12 In a) and b) the ammeter may be connected anywhere in the circuit provided it is in series but the voltmeter must be connected in parallel across component *R*.

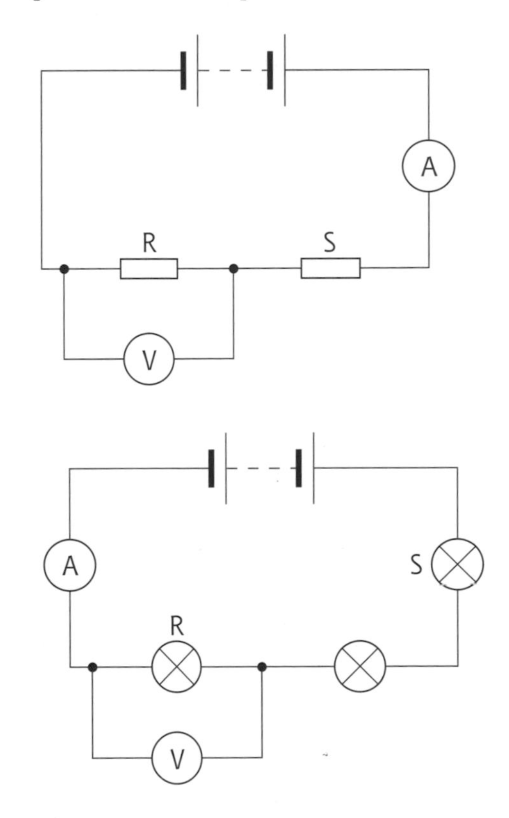

13 A = resistance, B = ohms, C = decreases, D = constant

14 a) 10 V, b) 12 V, c) 0.05 A, d) 5.75 A, e) 2 Ω, f) 400 Ω

15 a) 8 V, b) 0.5 A, c) 2 Ω

16 4 V; **17** 17.7 Ω; **18** 3 Ω

19 a) series, b) parallel, c) series, d) parallel

20 a) It is the same. b) Voltage of supply is equal to the sum of the voltages across each of the components.

21 a) Current drawn from supply is equal to the sum of the currents in each of the components.
b) It is the same.

22 a) Series circuit – current same at all points
$A_1 = A_2 = A_3 = A_4 = 4\,A$.
b) Parallel circuit – current splits up
$A_1 = 4.5\,A$, $A_2 = 3.0\,A$, $A_3 = 1.5\,A$,
$A_4 = 3.0\,A$.
c) Mixed series and parallel circuit.
series part, $A_2 = A_4 = 6\,A$
parallel part $A_1 = 6 - 4 = 2\,A$,
$A_3 = 6 - 5 = 1\,A$

23 a) Resistors in series – supply voltage splits up
i.e. $230 = 100 + V_1$ hence $V_1 = 130\,V$.
b) $V_2 = 4 + 6 = 10\,V$
c) Resistors in parallel – voltage same
$V_3 = 6\,V$
d) $12 = V_4 + 4$ hence $V_4 = 8\,V$

24 a) $A_1 = 1.5\,A$, $V_1 = 3\,V$
b) $A_2 = 5\,A$, $V_2 = 10\,V$
c) $A_3 = 2\,A$, $V_3 = 10\,V$

25 a) $25\,\Omega$, b) $33\,\Omega$, c) $210\,\Omega$

26 a) $8\,\Omega$, b) $12\,\Omega$, c) $15\,\Omega$

27 a) $5\,\Omega$, b) $12\,\Omega$, c) $12\,\Omega$

28 a) $20\,\Omega$, b) $49\,\Omega$, c) $8\,\Omega$

29 A = series

30 a) $30\,\Omega$, b) $0.2\,A$, c) i) $2\,V$, ii) $4\,V$

31 a) $V_1 = 4\,V$, $V_2 = 4\,V$,
b) $V_1 = 4\,V$, $V_2 = 8\,V$,
c) $V_1 = 1.67\,V$, $V_2 = 8.33\,V$

32 a) $V_2 = 3\,V$, $R_2 = 12\,\Omega$,
b) $V_2 = 4\,V$, $R_2 = 80\,\Omega$,
c) $V_2 = 4\,V$, $R_2 = 2000\,\Omega$

33 a) $V_S = 6\,V$, $R_1 = 30\,\Omega$,
b) $V_S = 100\,V$, $R_1 = 30\,\Omega$,
c) $V_S = 9\,V$, $R_1 = 125\,\Omega$

34 a) In series, b) For example a television and a hairdryer.

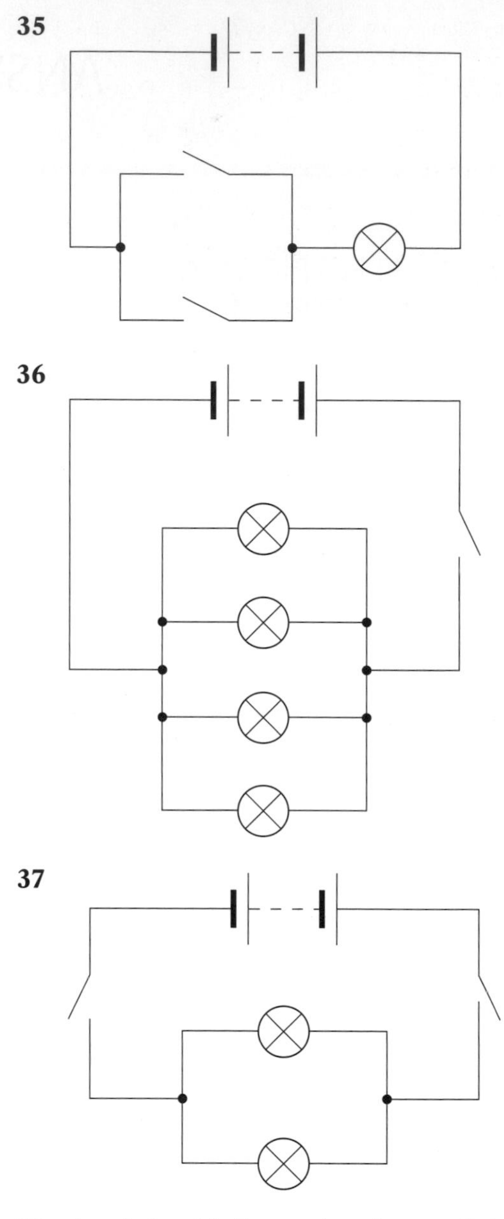

38 a) switch, b) fuse, c) resistor, d) battery, e) variable resistor, f) lamp, g) voltmeter, h) ammeter

39 a) In series. b) Touch X and Y together and lamp should light. c) i) Connect one end of fuse to X and other end to Y. ii) If lamp does not light then fuse is blown and if it lights then fuse is not blown.

1.2 Electrical energy

1 a) personal stereo, b) digital clock, c) electric fire, d) food mixer, e) television

2 a) electrical energy to heat, b) electrical energy to sound, c) electrical energy to heat, d) electrical energy to kinetic energy, e) electrical energy to sound

3 A = electrical, B = heat

4 A = joules, B = watts, C = divided, D = second, E = multiplied, F = voltage

5 a) 36 W, b) 1150 W, c) 20 A, d) 9.57 A, e) 15 V, f) 230 V

6 a) 60 W, b) 60 J

7 a) 46 W, b) 2300 W, c) 920 W

8 a) i) electrical energy to light, ii) filament or resistance wire of the lamp. b) 100 J

9 a) i) electrical energy to heat, ii) element, b) 4.6 A, c) 1058 W

10 a) $P = IV$ but $V = IR$ therefore $P = I(IR) = I^2R$

b) $P = IV$ but $I = \frac{V}{R}$ therefore $P = \frac{V}{R}V = \frac{V^2}{R}$

11 a) 1.44 W, b) 2645 W, c) 0.72 Ω, d) 35.3 Ω, e) 12 V, f) 10 V

12 7.2 W

13 A = one, B = direct, C = d.c., D = opposite, E = alternating, F = a.c.

14 A = 50, B = 230

15 A = less, B = peak, C = declared

16 a) The charges only move in one direction.
b) The charges move in one direction, then the other direction and so on i.e. to and fro.

17 Any value greater than 120 V

18 a) 1.2×10^{-10} W, b) 7.5×10^{4} Ω

19 12 V

20 electric kettle, electric cooker, electric toaster, etc.

1.3 Electromagnetism

1 a) The magnetic field surrounding the current-carrying wire causes the compass needles to move.
b) i) Compass needles return to their original positions as shown in diagram A.
ii) Compass needles point in the opposite direction i.e. anti-clockwise.

2 Increase voltage (current); wind more turns of wire around the nail.

3 Increase voltage (current); increase number of turns of wire; place an iron core through the turns of wire.

4 a) Diagrams A, C and E. Diagram D does not give a reading since both wires give a voltage in the same direction and this means that they cancel each other out.
b) Stronger magnetic field; more turns of wire passing through magnetic field; coils of wire moved faster through magnetic field.

5 The voltmeter shows a reading that returns to zero, then shows a reading of the opposite sign which returns to zero as the magnet makes one rotation.

6 A = magnetic, B = voltage, C = secondary, D = alternating

7 a) Require insulated covered wire and an iron core e.g. iron nail.
b) Wind a number of turns of wire onto the iron core. Then wind a second set of turns of wire to produce two separate coils of wire wrapped around the iron core.

8 a) V_P= 230 V, b) V_S = 20 V, c) N_P = 40 turns, d) N_S = 125 turns

9 i) a) V_S= 20 V, b) V_S = 6 V, c) V_S = 10 V, ii) The transformers are 100% efficient.

10 i) a) V_P = 20 V, b) V_P = 9 V, c) V_P = 230 V, ii) The transformers are 100% efficient.

11 i) a) N_S = 10 turns, b) N_S = 50 turns, c) N_S = 480 turns, ii) The transformers are 100% efficient.

12 i) a) N_S = 1840 turns, b) N_S = 480 turns, c) N_S = 1920 turns, ii) The transformers are 100% efficient.

13 a) V_P = 48 V, b) V_P = 10 V, c) I_P = 0.3 A, d) I_S = 1 A

14 a) I_P = 0.5 A, b) I_P = 1 A, c) I_P = 3.75 A

15 a) I_S = 0.2 A, b) I_S = 8 A, c) I_S = 0.14 A

16 a) V_S = 12 V, b) I = 3 A, c) I_P = 0.16 A

17 a) a.c., b) N_S = 120 turns, c) I_P = 0.013 A

18 a) 5 V, b) 1.25 A, c) 6.25 W

19 To minimise power lost (as heat) in the transmission cables.

20 Heat is produced in the coils and core; sound is produced as the core vibrates; some of the magnetic field is lost from the transformer.

21 A = step-up transformer, B = pylons, C = transmission cables, D = step-down transformer, E = homes, F = 25 000 V, G = 400 000 V, H = 230 V

22 a) 600 A, b) 25 Ω, c) 9×10^6 W, d) 231×10^6 W

1.4 Electronic components

1 a) motor, b) buzzer or lamp or LED, c) buzzer or lamp or LED

2 a) solar cell, b) LDR or solar cell, c) thermocouple or thermistor, d) microphone

3 a) sound to electrical energy, b) heat to electrical energy, c) light to electrical energy

4

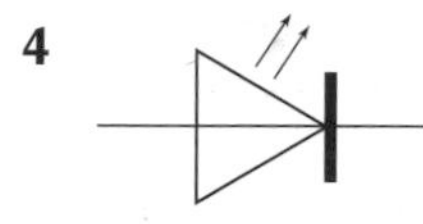

5 a) LED does not light, b) LED lights, c) LED does not light, (LED damaged by too large a current)

6 a) P = LED, Q = resistor, b) To prevent too large a current (or voltage) from damaging LED

7 a) series, b) 7.5 V, c) 1000 Ω

8 a) i) and ii) microphone – sound to electrical energy; solar cell – light to electrical energy; thermocouple – heat to electrical energy
b) i) and ii) lamp – electrical energy to heat and light; LED – electrical energy to light; loudspeaker – electrical energy to sound

9 a) Thermistor or thermocouple, b) microphone, c) LDR or solar cell

10 a)

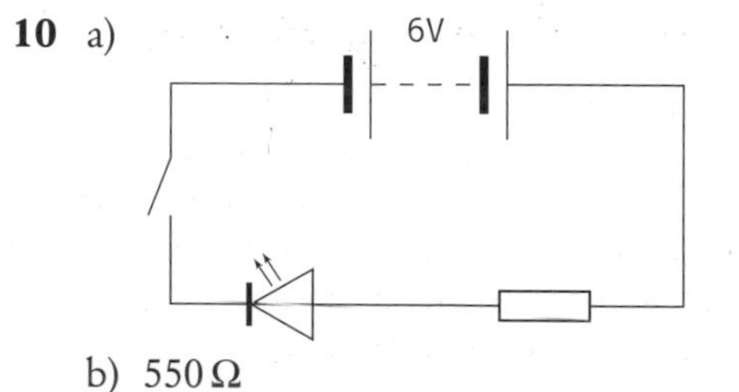

b) 550 Ω

11 a) Since temperature is the same the resistance of thermistor does not change so the reading on the ohmmeter remains at 20 000 Ω.
b) Since the temperature has increased the resistance of the thermistor will change. For most thermistors as the temperature increases their resistance decreases so the reading on the ohmmeter will usually be less than 20 000 Ω.

12 As the light intensity increases then the resistance of the LDR decreases so the reading on the ohmmeter will be less than 1500 Ω.

13 a) $R_{250\ \text{units}} = 2400\ \Omega$, $R_{500\ \text{units}} = 1200\ \Omega$, $R_{1000\ \text{units}} = 480\ \Omega$
b) Resistance of LDR decreases with increasing light intensity.
c)

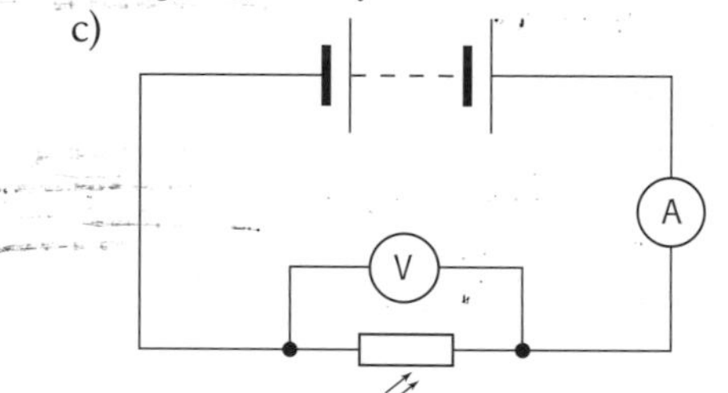

14 a) 40 Ω
b) Voltmeter reading remains at 10 V (equal to the supply voltage) but since $R_{\text{thermistor}}$ increases as temperature decreases then ammeter reading will decrease e.g. 0.20 A.

15 A = switch, B = voltage, C = non-conducting, D = conducting

16 a) W = thermistor, X = resistor, Y = transistor, Z = lamp
b) An electronic switch which switches on when the p.d. across X is equal to or above a certain value.

17 a) W = variable resistor, X = LDR, Y = MOSFET, Z = LED
b) An electronic switch which switches on when the p.d. across the LDR reaches a certain value.

18 a) On (conducting)
b) As the control on the potentiometer is moved towards Y, the resistance between Y and Z decreases. The voltage between Y and Z decreases and when it decreases below a certain value the transistor switches off (non-conducting) and the LED goes out.

19 a) i) 0 V, ii) Off (non-conducting), iii) Unlit
b) Reading on voltmeter is 3 V so transistor switches on and LED lights.

20 a) Circuit A – to switch on a lamp when it gets dark; Circuit B – the LED lights when an element of a cooker is hot.
b) The variable resistor allows the circuit to be adjusted to switch the transistor on for different conditions (e.g. a different light level or a different temperature).

21 a) As the temperature increases, the resistance of the thermistor decreases, so there will be less voltage across the thermistor. This means that there is more voltage across the resistor and when it is greater than or equal to a certain value the MOSFET switches on and the buzzer sounds.
b) Same diagram as given in the question but with the positions of the thermistor and resistor R interchanged.

22 In moving to the dimly lit room the resistance of the LDR increases. The voltage across the LDR increases and when it is greater than or equal to a certain value the transistor switches on and the LED lights.

23 a)

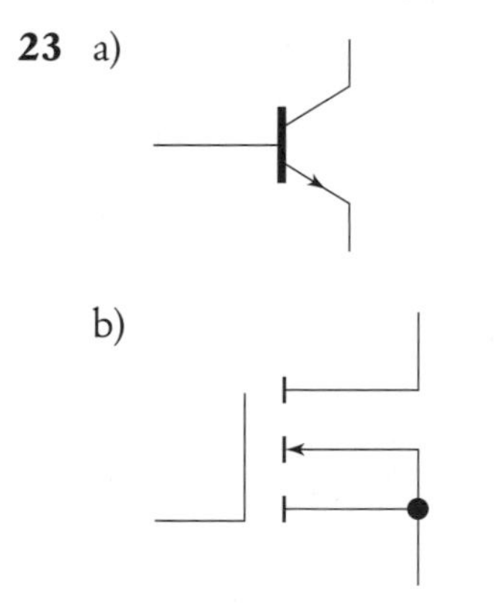

b)

24 baby alarm; hi-fi; radio; television.

25 a) To make the electrical signal larger.
b) Frequency of the input signal = frequency of the output signal.
c) Amplitude of the output signal > amplitude of the input signal.

26 a) 400, b) 2500, c) 24 V, d) 23.1 V, e) 0.2 V, f) 1.2×10^{-3} V

27 200; **28** 1.5 V; **29** 2.67×10^{-4} V

30 Connect voltmeter V_1 across the input to the amplifier. Connect another voltmeter V_2 across the output from the amplifier. Note readings on both voltmeters.

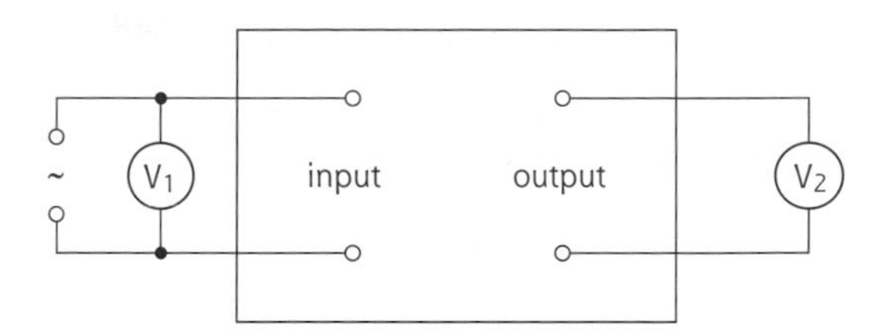

Calculate the voltage gain using the equation:

$$V_{gain} = \frac{V_{output}}{V_{input}} = \frac{\text{reading on } V_2}{\text{reading on } V_1}$$

Exam-style questions

1 a) Potential divider
b)

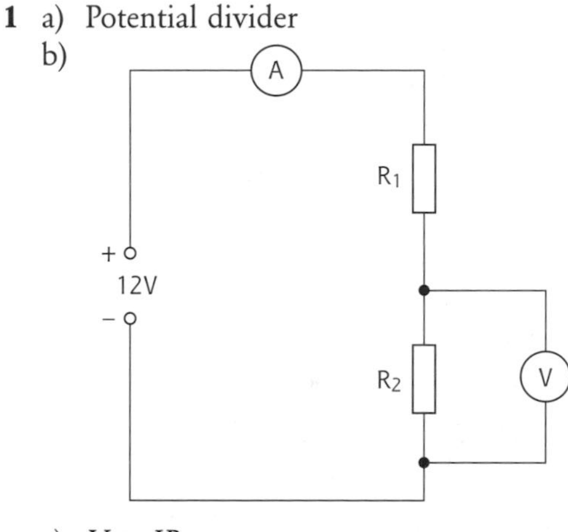

c) $V = IR_2$
$3 = 0.25 \times R_2$
$R_2 = \frac{3}{0.25} = 12\,\Omega$
d) 36 Ω

2 a) parallel, b) 3 Ω, c) 4 A

3 a) i) loudspeaker and LED
ii)

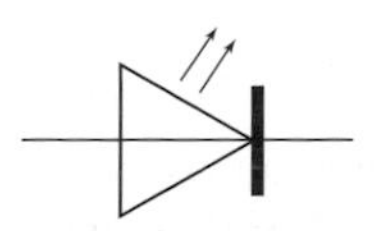

iii) sound to electrical energy
b) X = Light dependent resistor (LDR), Y = transistor
c) As room gets darker then resistance of LDR increases and the voltage across the LDR increases. When voltage across LDR is equal to or above a certain value the transistor switches on and the lamp lights.

4 a) i) 3×10^{-3} A, ii) 1000 Ω, iii) From graph temperature = 10 °C
b) As the temperature decreases the resistance of the thermistor increases and the voltage across the thermistor increases. When the voltage across the thermistor is equal to or above a certain value the transistor switches on and the lamp lights.

5 a) If one lamp breaks in the set then the other lamp remains lit.
b) i) 24 Ω, ii) 6 Ω, c) i) 4 A, ii) 5 A

6 a) Y, b) 730 Ω

7 a) 35, b) 500 Hz

8 a) alternating current
b) electrons move in one direction, then in the opposite direction and then back again and so on.
c) 5.5 A, d) 1650 C

9 a) i) The voltmeter shows a reading which returns to zero, then a reading of the opposite sign which returns to zero as the magnet makes each oscillation.
ii) Increase number of turns of wire on the coil; use a magnet with a stronger magnetic field; make magnet move faster through the turns of wire on the coil.
b) i) To change the size of an a.c. voltage.
ii) A) 3600 turns, B) Transformer is 100% efficient.
iii) $P_{input} = IV = 1.7 \times 16 = 27.2$ W
$P_{lost} = P_{input} - P_{output} = 27.2 - 24$
= 3.2 W i.e. 3.2 J of electrical energy are lost each second.
iv) Sound is produced due to vibration of coils and core; heat is produced in coils; magnetic field is lost from the core.

Chapter 2 MECHANICS AND HEAT

2.1 Kinematics

1 Note initial distance (odometer) reading before journey commences. Note final distance (odometer) reading after completing journey. Start stopwatch when journey starts and stop stopwatch when journey is completed. Note time taken for journey. Calculate distance travelled from final distance reading minus initial distance reading.

Calculate average speed $= \dfrac{\text{distance travelled}}{\text{time taken}}$

2 a) 40 m/s, b) 533 m/s, c) 9×10^6 m, d) 15 m, e) 200 s, f) 8 s

3 4 m/s; **4** 1500 m; **5** 12.5 s; **6** 25 m/s;
7 3750 s; **8** 60 mm

9 Mark a line on the road. Start stopwatch when the front wheel of the bicycle passes this line and stop stopwatch when rear of bicycle passes the line. Measure length of bicycle.

Calculate speed $= \dfrac{\text{length of bicycle}}{\text{time on stopwatch}}$

10 a) Average speed is the total distance travelled divided by the total time taken for the journey. Instantaneous speed is the speed of an object at a certain time during the journey.
b) A roller coaster; an object accelerating down a hill.

11 a) A vector quantity has magnitude and direction.
b) displacement or velocity

12 a) 10 km, b) 7.6 km at 67° N of E

13 a) 5 km at 53° E of S, b) 3.6 km at 34° E of S, c) 4.2 km at 45° E of N, d) 4.2 km at 45° S of E

14 121 m/s at 8.5° E of N

15 2.6 m/s at 18° W of S

16 13 m/s at 18° downstream from direction rower is rowing

17 6.8 m/s at 54° S of E

18 a) Speed is a scalar quantity and is the distance travelled by an object in 1 s.
b) Acceleration is a vector quantity and is the change in velocity in 1 s.

19 a) The velocity of the object increases by 2 m/s every second.
b) The velocity of the object decreases by 4 m/s every second.

20 a) 8 m/s², b) 7 m/s², c) 4 s, d) 40 s, e) 205 m/s, f) 51.2 m/s

21 4 miles per hour per second

22 a) 4 m/s², b) 4 m/s², c) −4 m/s², d) −6.25 m/s², e) 32.5 m/s, f) 82.3 m/s, g) 5 m/s, h) 31 m/s, i) 2.5 s, j) 2.2 s

23 12 m/s; **24** 2 m/s²; **25** 0.17 m/s²; **26** −0.4 m/s²; **27** 4 s; **28** 1.25 m/s²; **29** 7.5 s; **30** 0.8 s

31

32

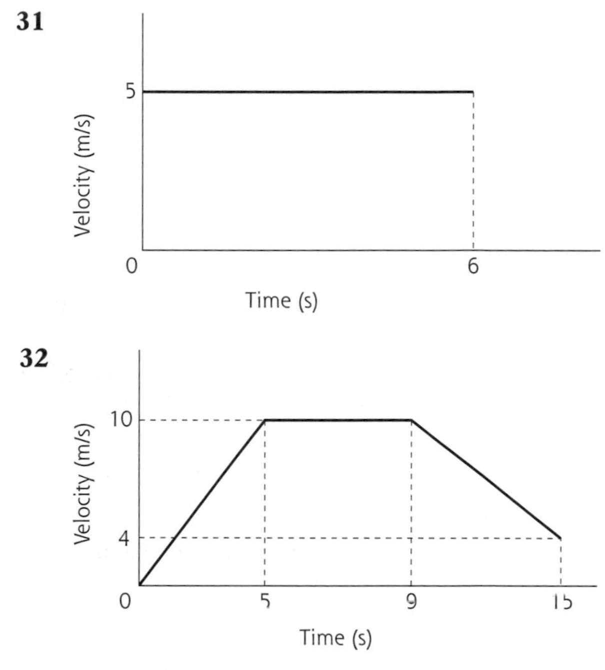

33 a) constant speed, b) constant acceleration from rest, c) constant negative acceleration (constant deceleration), d) constant acceleration not from rest

34 a) 1.25 m/s², b) −1.4 m/s², c) 0.4 m/s², d) −1.5 m/s²

35 a) 60 m, 2.5 m/s², b) 54 m, −3 m/s², c) 96 m, 4 m/s², −2.4 m/s², d) 77.5 m, 1 m/s², −3 m/s²

36 a) i) constant acceleration, ii) constant acceleration, b) i) 10 m/s², ii) 10 m/s², c) 0 m, d) 1.8 m

37 a) i) constant acceleration, ii) constant velocity, iii) constant deceleration
b) i) 2 m/s², ii) 0, iii) −1.25 m/s²,
c) 560 m, d) 13.7 m/s

38 a) In method A, the student has to judge when the front of the trolley reaches point X and so start the stopwatch and judge when the rear of the trolley reaches point X to stop the stopwatch. In method B, the timer starts when the trolley breaks the beam at point X and stops when the light beam is restored.
b) Method B

2.2 Forces

1 Change the speed, direction and/or shape of an object.

2 Attach Newton balance to the object. Pull on Newton balance. The reading on the Newton balance is now equal to the force applied to the object.

3 Gravitational field strength, g, is 10 N/kg

4 a) 25 N, b) 170 N, c) 9 kg, d) 22 kg, e) 80 N, f) 288 N

5 70 N; **6** 0.6 kg

7 a) 96 N, b) The moon exerts a force of 1.6 N on each kg of an object, c) 600 N,
d) i) 60 kg
ii) Mass is the quantity of matter and is not dependent on location and so will remain the same.

8 1140 N

9 Mass is the quantity of matter and is measured in kg. Weight is the pull of the Earth (or a planet) on an object. It is a force and is measured in N.

10 a) north, b) south

11 a) To stop a car or a bicycle by using the brakes. b) To allow a car to accelerate by releasing the brakes or lubricating a surface.

12 a) engine force and frictional force
b) They are equal in size but in the opposite direction.

13 a) forward thrust and air resistance,
b) 280 kN

14

Direction of travel

Force of friction

Engine force

15 a) 12 N, b) i) 12 N
ii) Upwards. From Newton's First Law, since the lamp shade is stationary, then there must be balanced forces acting on it. One force is the weight acting downwards. Hence, the tension must be equal in size but in the opposite direction to the weight.

16 a) The weight of the parachutist is equal in size but in the opposite direction to air resistance.
b) The weight of the clock is equal in size but in the opposite direction to the force (reaction) of the table on the clock.
c) Pedalling force is equal in size but in the opposite direction to the frictional force.

17 When the brakes are applied, then a force acts in the opposite direction of motion on the car or bus, which decelerates. This force does not act on the person and so they would continue to travel at constant speed in a straight line. However, the seat belt applies a force, in the same direction as the braking force, which decelerates the person.

18 a) 4 N to the right, b) 4 N to the left, c) 0 N, d) 30 N downwards

19 a) 90 N, b) 2 N, c) 4 m/s^2, d) 30 m/s^2, e) 8 kg, f) 26.7 kg

20 3000 N; **21** -1.65 m/s^2; **22** 8×10^4 kg

23 a) 15 N, b) 12 N, c) 8 m/s^2, d) 3 m/s^2, e) 24 kg, f) 5 kg

24 a) 1.5 N, b) 0.5 N

25 a) 3240 N, b) 3740 N

26 a) As mass decreases, the acceleration increases.
b) As the unbalanced force decreases, the acceleration decreases.

27 a) constant speed, b) constant acceleration

28 a) 81 m, b) 45 m/s, c) $\frac{(u + v)}{2} = \frac{(0 + 45)}{2} = 22.5$ m/s, d) 101 m

29 a) 5 m, b) Height rocket falls is 0.2 m so rocket does not hit target.

30 a) 6 m/s, b) 3 m/s, c) 1.8 m, d) 12 m/s

31 a) 4 m, b) 1.25 m, c) 8.9 m/s at 26.6° below the horizontal.
d) No differences. Since there are no forces acting horizontally then the horizontal speed will remain constant. Vertically all objects accelerate at the same rate of 10 m/s^2 when no air resistance is present so the vertical speed will increase at 10 m/s every second.

2.3 Momentum and energy

1 4.8 m/s; **2** 4.9 m/s; **3** 4 kg; **4** 7.5 m/s; **5** 2.5 m/s; **6** 10.3 m/s; **7** 9.8 m/s; **8** 10.7 m/s; **9** 11.3 m/s; **10** 20 m/s

11 a) chemical energy to kinetic energy, b) chemical energy to heat, c) kinetic energy to heat

12 a) chemical energy to gravitational potential energy and heat, b) gravitational potential energy to heat

13 a) gravitational potential energy to kinetic energy, b) kinetic energy to elastic potential energy to kinetic energy, c) kinetic energy to gravitational potential energy

14 a) 175 J, b) 135 J, c) 20 m, d) 20 m, e) 60 N, f) 40 N

15 150 J; **16** 75 N; **17** 30 m; **18** 250 m;
19 240 N

20 a) 3000 N, b) 1.5×10^6 J

21 a) 2400 W, b) 5 W, c) 5.28×10^5 J, d) 9.38×10^5 J, e) 25 s, f) 60 s

22 140 W; **23** 1440 J; **24** 5 s

25 a) 6 J, b) 160 J, c) 5 m/s, d) 7 m/s, e) 80 kg, f) 2.5 kg

26 45 J; **27** 2.16 kg; **28** 2 m/s;
29 1.2×10^{-3} J; **30** 0.2 kg; **31** 30 m/s

32 a) 175 J, b) 3.5 J, c) 4 kg, d) 2.67 kg, e) 6 m, f) 16 m

33 90 J; **34** 1.2 m; **35** 2.5 kg; **36** 692 J

37 a) 75 J, b) 75 J, c) 10 m/s

38 a) 1440 J, b) 1440 J, c) 3.2 m

39 a) 4×10^4 J, b) 1333 W

40 a) 1.75×10^6 J, b) 3.5×10^4 W

41 a) 80%, b) 50%, c) 225 J, d) 157 J, e) 4.8 MJ, f) 400 kJ

42 a) 80%, b) 50%, c) 225 W, d) 157 W, e) 4.8 MW, f) 400 kW

43 47.6%

44 a) 7.5×10^5 J, b) 6×10^5 J

45 a) 20 W, b) 21 600 J

46 a) 4000 W, b) 8%

2.4 Heat

1 A = temperature, B = celsius, C = heat, D = conduction

2 Aluminium, since aluminium has a larger specific heat capacity than copper.

3 a) 6.27×10^4 J, b) 1.00×10^5 J, c) 5.02×10^5 J

4 a) 1500 J, b) 200 J/kg °C, c) 0.5 kg, d) 50 °C

5 9020 J; **6** 1.42×10^5 J

7 a) 9650 J, b) 7720 J, c) 80%

8 334 s; **9** 2000 J/kg °C

10 A = $cm\Delta T$, B = specific heat capacity, C = ml, D = fusion, E = vaporisation, F = temperature

11 a) 500 000 J, b) 5 kg, c) 1.25×10^5 J kg^{-1}, d) 67 800 J, e) 0.7 kg

12 a) 334 000 J, b) 668 000 J, c) 1.67×10^6 J, d) 133 600 J

13 a) 2.26×10^6 J, b) 4.52×10^6 J, c) 1.13×10^7 J, d) 9.04×10^5 J

14 0.05 kg; **15** 1.08×10^{-2} kg

16 a) 2.93×10^4 J, b) i) 2.26×10^5 J, ii) 2.55×10^5 J
c) Approximately 10 times as much energy is given to the skin during a scald from steam than a scald from boiling water. This means that the burn from the steam is more severe than that from boiling water.

17 200 000 J/kg; **18** b) 1200 s

19 Refrigerator, freezer, or ice pack

Exam-style questions

1 a) Donna uses the measuring tape to measure the length of the straight road. When Jenny enters the straight road Donna starts her stopwatch. When Jenny leaves the straight road Donna stops her stopwatch.

$$\text{Average speed} = \frac{\text{length of road}}{\text{time recorded on stopwatch}}$$

b) i) 150 N, ii) 120 J

2 a) 1200 N – constant speed so balanced forces
b) i) -0.0048 m/s^2, ii) 312.5 s
iii)

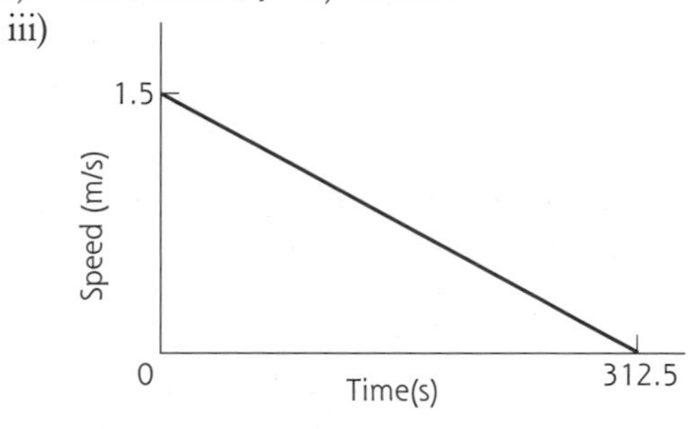

iv) distance = area under graph
$= \frac{1}{2} \times 312.5 \times 1.5$
= 234 m

3 a) 1 250 J, b) i) 83.3 W
ii) Some of the input energy to the motor is changed into heat and sound in the motor and not raising the bales.

4 a) 1.0 s, b) -1.43 m/s^2
c) Distance travelled = 600 m so train stops 20 m short of the signal.

5 a) 0.6 J, b) i) gravitational potential energy to kinetic energy
ii) assuming no energy transfer to surroundings, change in E_K = change in gravitation E_P.

6 a) 2700 J, b) 11 m/s, c) 180 m

7 a) i) 0.15 m/s^2, ii) 720 m, iii) 4 m/s
b) i) unbalanced boat is accelerating,
ii) balanced boat is travelling at constant speed.

8 a) 13.4 m/s, 26.6° W of S, b) 120 N, balanced forces since moving at constant speed, c) 8.2 m/s

9 a) 5.25 m/s, b) i) 22 m, ii) 42 m/s, iii) 88.2 m

10 a) i) constant acceleration, ii) constant acceleration, b) 1.25 m, c) 0.45 m
d) During the collision with the ground some of the ball's kinetic energy is changed into sound. The ball rebounds with less kinetic energy and so rises to a lower height.
e) 0.5 s, since the velocity changes sign showing that the ball is moving in the opposite direction.

11 a) 517 J/kg °C, b) steel

12 a) 83.6 s
b) The energy supplied by the kettle is used to increase the temperature of not only the water but other things such as the element, the materials that the kettle is made of and some of the air surrounding the kettle.
c) 0.088 kg

13 a) 9.94×10^6 J, b) 26.3 °C
c) The energy supplied by the immersion heater is used to increase the temperature of not only the water but the materials that the tank is made of and some of the air surrounding the tank.

14 a) 0.02 A, b) 69 J, c) 2.7 J, d) 3.91%

Chapter 3 WAVES AND OPTICS

3.1 Waves

1 a) Speed of light is faster than the speed of sound.
b) 340 m/s, c) Storm is further away

2 a) When they hear the sound. b) 329 m/s
c) The time will be less and the reaction time will now be more important.

3 2550 m

4 a) 1500 m/s, b) 330 m/s, c) 3.1 m, d) 2211 m, e) 5.5 s, f) 0.3 s

5 1088 m

6 a) 1260 m, b) 560 m
c) Energy is absorbed by the water and changed into heat during transmission.

7 2 m; **8** Energy

9 a) i) 0.5 m, ii) 2m, b) i) 0.25 m, ii) 1 m, c) i) 0.375 m, ii) 0.6 m, d) i) 0.85 m, ii) 2 m, e) i) 0.125 m, ii) 0.25 m, f) i) 0.1 m, ii) 0.36 m

10 a) There are 50 waves every second.
b) The length of one complete wave is 0.3 m.

11 a) 2.5 Hz, b) 8.4 Hz, c) 7 Hz, d) 240 Hz, e) 40 Hz, f) 1.25×10^6 Hz, g) 2.4×10^6 Hz, h) 4 000 Hz

12 a) 30.8 m/s, b) 36 m/s, c) 1.5 m, d) 80 m, e) 133 Hz, f) 250 Hz

13 30 m/s; **14** 6×10^{-4} m; **15** 120 000 Hz

16 Particles of the medium vibrate at right angles to the direction of the energy transfer.

17 Particles of the medium vibrate parallel to the direction the wave is travelling in (the direction of the energy transfer).

18 3×10^8 m/s

19 a) 3.15×10^8 m/s, b) 1505 m/s, c) 2.1×10^{-7} m, d) 6×10^{14} Hz, e) 2.7×10^{14} Hz, f) 4.4×10^{-4} m

20 A = Ultraviolet, B = infra red, C = TV

21 0.017 m

22 a) photographic plate, b) fluorescent substance, c) aerial and receiver

23 a) infra red, b) larger

3.2 Reflection

1 a) Normal
b)

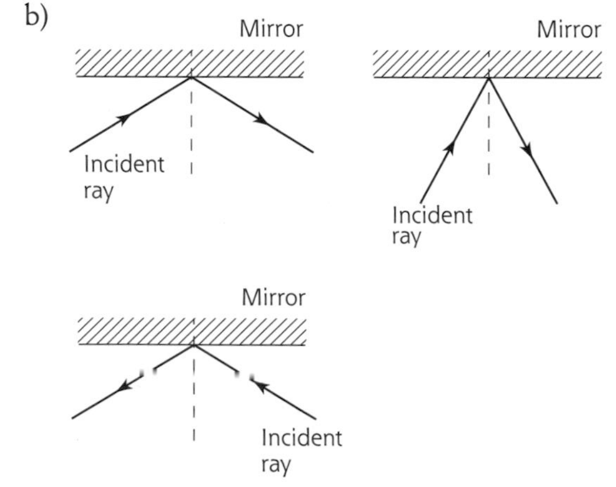

2 a)

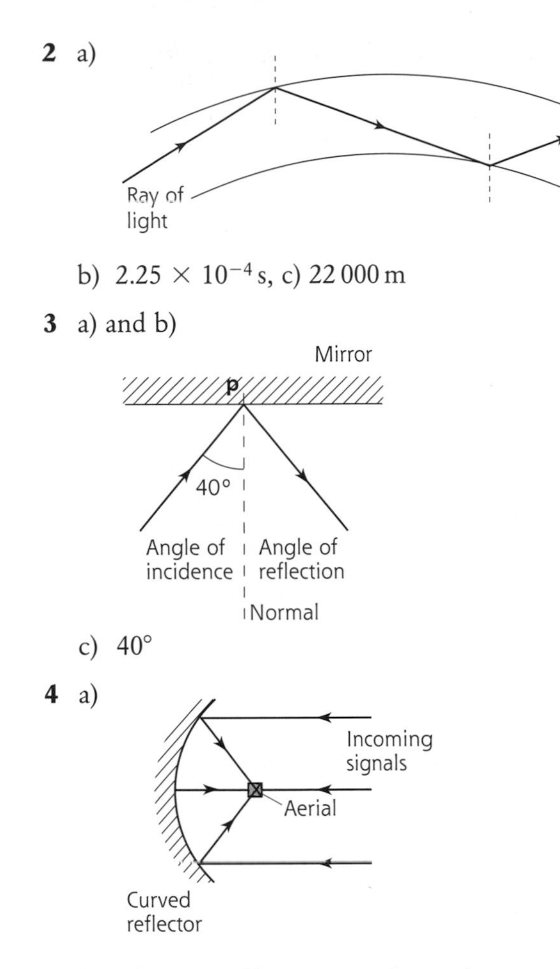

b) 2.25×10^{-4} s, c) 22 000 m

3 a) and b)

c) 40°

4 a)

b) The rays will come to a focus closer to the dish so the signal received at the aerial will be weaker.

5 a)

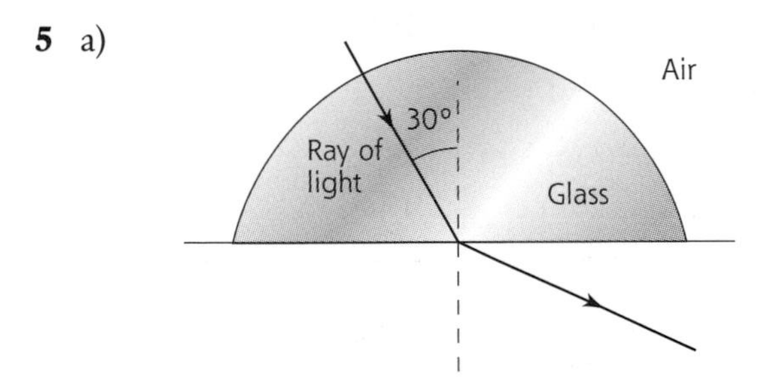

b) The ray will be totally internally reflected.

6 a) The critical angle is the angle at which the emerging ray passes along the surface of the block of glass.

b)

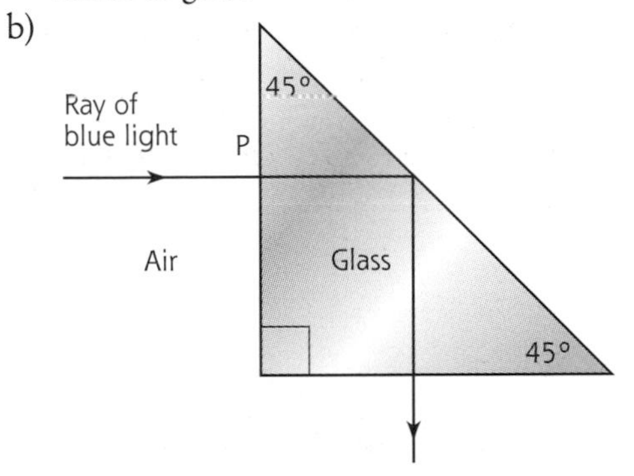

7 a) electrical cables, b) light, c) electrical to light to electrical to sound, d) higher rates of information sent/cost of fibres cheaper/no interference

3.3 Refraction

1 a) Refraction is the passage of waves from one substance to another and so the speed of the waves changes

b)

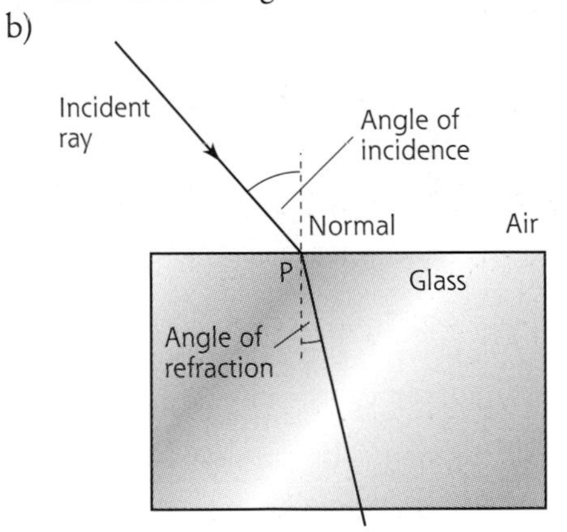

c) any value less than 40°

2 a)

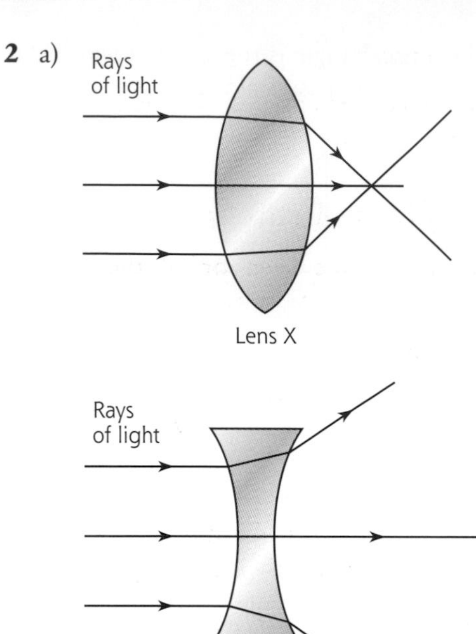

b) Parallel rays will come to a focus 100 mm from the centre of the lens.

3 b) real and magnified

4 b) 53 mm

6 a) 86 mm, real and diminished, b) 240 mm, real and magnified, c) 60 mm, virtual and magnified

7 a) Long sight due to power of lens being positive and must be a convex lens.
b) 0.4 m

8 a) 6.7, b) −5.0, c) 400 mm, convex, d) 250 mm, concave, e) −2, f) 3.3

9 a) short sighted, b) concave lens, c) −1.7 D

10 a) Short sight since power of lens is negative which is a concave lens.
b) 0.625 m
c) Right eye, since power less. This means focal length is greater.

Exam-style questions

1 b) i) real ii) magnified; **2** 100 mm

3 a) Gallium arsenide, Neodymium YAG and carbon dioxide
b) 4.7×10^{14} Hz, c) i) excimer,
ii) fluorescent materials

4 a) i) short sight, ii) concave, b) 0.31 m
c) Focus a distant object through the lens on to the screen. Measure distance from lens to screen. This is the focal length.

5 a) infra red, b) greater, c) possibility of skin cancer

6 a) i) refraction
ii) The speed of light changes when it enters the glass.
b) i)

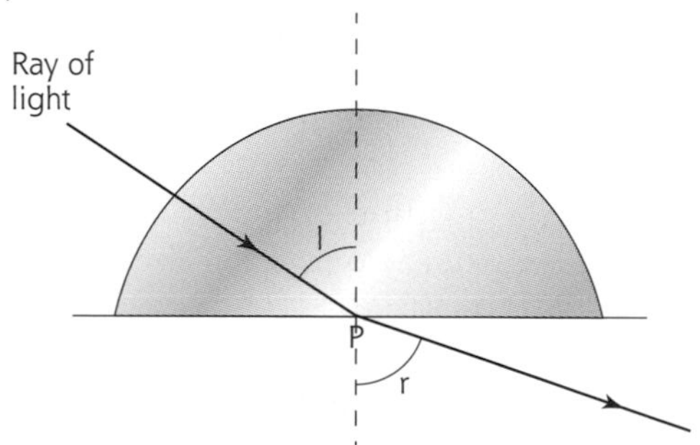

ii) l = angle of incidence, r = angle of refraction
iii) Angle of refraction gets bigger and bigger. When angle of incidence in glass is greater than 42° the light is totally internally reflected.

Chapter 4 RADIOACTIVITY

4.1 Ionising radiations

1 a) beta and gamma, b) gamma, c) gamma

2 a) electron, b) neutron, c) proton

3 a) alpha, b) gamma, c) beta

4 a) Addition or removal of an electron to produce a charged particle.
b) i) Alpha radiation
ii) Large number of ions in a small volume (large ionisation density) and this has a big effect on the living cells inside the body.

5 The film blackens with radiation and the degree of blackening increases with the amount of radiation detected.

6 X – alpha, beta and gamma since there is no absorber present.
Y – gamma since alpha and beta are absorbed by the aluminium.

7 Kill cancer cells.

8 Use as tracers to detect illness.

9 a) Passes outside the body to be detected.
b) Left kidney
c) More radiation retained by this kidney indicating failure.

10 a) easily absorbed, b) between 2 cm and 5 cm
c) An air gap of between 2 cm and 5 cm will not absorb the alpha particles. If smoke enters the gap then the alpha particles will be absorbed and the alarm will go off.

11 a) Beta since different thicknesses of cardboard will absorb different amounts of beta radiation.
b) As thickness of card increases, count rate decreases.
c) Put a detector next to card and show that count rate is similar to the background count rate.

12 Short range and high ionisation density so has a big effect on the cells of the tumour and kills them.

4.2 Dosimetry

1 a) 5 million nuclei disintegrate every second, b) decreases

2 a) 2×10^4 Bq, b) 700 Bq, c) 1.75 s, d) 1 s, e) 2.16×10^9, f) 2.7×10^6

3 40 Bq; **4** 1.35×10^9; **5** 30 s; **6** 3×10^9

7 a) 0.06 Gy, b) 5×10^4 Gy, c) 5×10^{-2} J, d) 4×10^{-4} J, e) 50 kg, f) 50 kg

8 0.0275 J; **9** 0.65 kg; **10** 4×10^{-3} Gy

11 Absorbed dose, type of radiation, type of tissue exposed.

12 a) 1×10^{-3} Sv, b) 1×10^{-4} Sv, c) 2.25×10^{-7} Gy, d) 2.5×10^{-3} Gy, e) 20, f) 1

13 0.50 Sv; **14** slow neutrons

15 2×10^{-5} Gy; **16** 7×10^{-3} J;

17 2.5×10^{-5} Sv; **18** 2.25×10^{-4} Sv;

19 2.01×10^{-3} Sv

20 a) 6×10^{-5} Sv, b) 6×10^{-6} Gy for alpha; 6×10^{-6} Gy for fast neutrons

21 3; slow neutrons

22 a) cosmic rays, radon gas, b) medical treatment, nuclear fallout

4.3 Half-life and safety

1 6 hours is the time taken for half the radioactive nuclei to decay.

2 275 Bq; **3** 6 hours

4 a) To prevent radioactivity reaching anyone.
b) 8 hours

5 50 MBq; **6** 156 kBq

7 a) Bq, b) decreases with time

8 6 hours; **9** 15.6 kBq; **10** 8.75 mg; **11** 480 kBq; **12** 2.7 days; **13** 24 days

14 a) 192; 152; 120; 96; 76; 60; 48,
b) 15 minutes

15 12 hours

16 a) 8 days
b) Assuming source count rate is the same as the activity of the source and the final count rate will be more affected by the varying background count rate.
c) Background will vary so repeated measurements taken and an average taken.

17 a) Source Y
b) Source X has too short a half-life, source Z has too long a half life and patient may be exposed to too much radiation.

18 Do not eat when working with sources; do not point the source at anyone; no one under 16 allowed to handle sources.

4.4 Nuclear reactors

1 No greenhouse gases, plentiful supplies of uranium. For a given mass more energy is released compared to fossil fuels.

2 Disposal of waste difficult. Risk of accident can cause damage to humans.

3 a) fission
b) Breaks into smaller nuclei and releases further neutrons which can strike other nuclei and release energy
c) kinetic energy

4 a) A neutron strikes a nucleus which splits and releases further neutrons which can split other nuclei and the process continues.
b) To strike other nuclei and continue the process.

5 a) Contain material which can undergo fission.
b) The material will no longer undergo fission.
c) Material may be released accidentally and cause damage; high radioactivity could damage humans in handling the material.

6 To prevent any radiation escaping and causing any damage.

7 a) X = control rods, Y = Moderator, Z = fuel rods
b) X = absorbs neutrons and so number of fission reactions can be controlled.
Y = slows down neutrons and so increases the chance of fission taking place.
Z = contain enriched uranium so that fission can occur.

8 a) X has water to allow for cooling of waste and is away from centres of population in case of accidents.
b) Marshy ground may allow waste to sink over long period of time and is near to centre of population which could harm a lot of people in the event of an accident.

Exam style questions

1 a) X – alpha, Y – gamma
b) Gain or loss of an electron to produce a charged particle.
c) X – alpha radiation

2 a) Every 6 hours half the radioactive nuclei of the source decays.
b) Only gamma radiation can get out from inside the body and be detected.
c) The tumour is always receiving radiation but the healthy tissue receives a lower amount.

3 a) 1 kBq, b) i) 0.05 Gy, ii) 0.05 Sv, c) Lead

4 a) 40%, b) fission
c) Keep as far from people as possible; store in lead or concrete containers.
d) To allow radiation to be absorbed by water until a suitable level of radiation is reached to allow for further treatment.

FORMULAE SHEET

$d = vt$

$s = vt$

$a = \frac{\Delta v}{t}$

$a = \frac{v - u}{t}$

$W = mg$

$F = ma$

$P = mv$

$E_w = Fd$

$E_p = mgh$

$E_k = \frac{1}{2} mv^2$

$P = \frac{E}{t}$

$percentage\ efficiency = \frac{useful\ E_o}{E_i} \times 100$

$percentage\ efficiency = \frac{useful\ P_o}{P_i} \times 100$

$E_h = cm\Delta T$

$E_h = ml$

$Q = It$

$V = IR$

$R_T = R_1 + R_2 + \ldots$

$\frac{1}{R_T} = \frac{1}{R_1} + \frac{1}{R_2} + \ldots$

$V_2 = \left(\frac{R_2}{R_1 + R_2}\right) V_s$

$\frac{V_1}{V_2} = \frac{R_1}{R_2}$

$P = IV$

$P = I^2R$

$P = \frac{V^2}{R}$

$\frac{n_s}{n_p} = \frac{V_s}{V_p} = \frac{I_p}{I_s}$

$V_{gain} = \frac{V_o}{V_i}$

$P_{gain} = \frac{P_o}{P_i}$

$v = f\lambda$

$P = \frac{1}{f}$

$A = \frac{N}{t}$

$D = \frac{E}{m}$

$H = Dw_R$

DATA SHEET

Speed of light in materials

Material	Speed in m/s
Air	3.0×10^8
Carbon dioxide	3.0×10^8
Diamond	1.2×10^8
Glass	2.0×10^8
Glycerol	2.1×10^8
Water	2.3×10^8

Gravitational field strengths

	Gravitational field strength on the surface in N/kg
Earth	10
Jupiter	26
Mars	4
Mercury	4
Moon	1.6
Neptune	12
Saturn	11
Sun	270
Venus	9

Specific latent heat of fusion of materials

Material	Specific latent heat of fusion in J/kg
Alcohol	0.99×10^5
Aluminium	3.95×10^5
Carbon dioxide	1.80×10^5
Copper	2.05×10^5
Iron	2.67×10^5
Lead	0.25×10^5
Water	3.34×10^5

Specific latent heat of vaporisation of materials

Material	Specific latent heat of vaporisation in J/kg
Alcohol	11.2×10^5
Carbon dioxide	3.77×10^5
Glycerol	8.30×10^5
Turpentine	2.90×10^5
Water	22.6×10^5

Speed of sound in materials

Material	Speed in m/s
Aluminum	5200
Air	340
Bone	4100
Carbon dioxide	270
Glycerol	1900
Muscle	1600
Steel	5200
Tissue	1500
Water	1500

Specific heat capacity of materials

Material	Specific heat capacity in J/kg °C
Alcohol	2350
Aluminium	902
Copper	386
Glass	500
Ice	2100
Iron	480
Lead	128
Oil	2130
Water	4180

Melting and boiling points of materials

Material	Melting point in °C	Boiling point in °C
Alcohol	−98	65
Aluminium	660	2470
Copper	1077	2567
Glycerol	18	290
Lead	328	1737
Iron	1537	2747

Radiation weighting factors

Type of radiation	Radiation weighting factor
alpha	20
beta	1
fast neutrons	10
gamma	1
slow neutrons	3